Studies on the Structure of Time

Studies on the Structure of Time

From Physics to Psycho(patho)logy

Edited by

Rosolino Buccheri
IFCAI-CNR
Palermo, Italy

Vito Di Gesù
University of Palermo
Palermo, Italy

and

Metod Saniga
Astronomical Institute
Slovak Academy of Sciences
Tatranská Lomnica
Slovak Republic

Kluwer Academic / Plenum Publishers
New York, Boston, Dordrecht, London, Moscow

Proceedings of the International Interdisciplinary Workshop on Studies on the Structure of Time: From Physics to Psycho(patho)logy, held November 23–24, 1999, in Palermo, Italy

ISBN 0-306-46439-X

233 Spring Street, New York, N.Y. 10013

http://www.wkap.nl

10 9 8 7 6 5 4 3 2 1

A C.I.P. record for this book is available from the Library of Congress

Printed in the United States of America

Preface

The problem of understanding the nature of time and the reason of its apparent flow has always captured the interest of Philosophers, Scientists and Theologists. Nowadays, especially after the great achievements in science, with special reference to the evolutionary processes in biology, in chemistry and in geology, many scientific groups in the world are involved in the study of time, approaching the problem from different points of view.

In organising the present workshop we wanted to focalise our attention in the discrepancy between the concept of time as described in physics and the way time is perceived by humans. We were challenged, in particular, by the search of relations between the two concepts of time and of their possible description using the language of science. Our starting point was the study of the so-called *altered states of consciousness*, those states of mind where the perception of time is very much different from "normality".

We were aware that the achievement of sound results in this study required necessarily the contribution of knowledge coming from several different disciplines; therefore psychologists, physicists, physicians, mathematicians and philosophers were invited to express their point of view and describe the results of their research connected with this topic.

The proposal of such a workshop sounded successful so that, just after the first announcement, we received a large number of requests of information and participation from scientists of different cultural areas. Moreover, many people, who could not be present at the meeting, are still asking for the release of these proceedings, thus confirming the constant live interest in this subject.

We hope that the idea contained in this book will stimulate to consider new approaches of investigation for the solution of the enigma of time. Moreover, taking into account the intense and fruitful exchange of information we had in the course of the meeting, is our aim to promote a permanent scientific debate and discussion among all participants.

It is a pleasure to thank all those who contributed to the organisation and to the success of the meeting. We thank in particular the Italian National Research Council – Area di Ricerca di Palermo and the Centro Interdipartimentale di Tecnologie della Conoscenza of the Palermo University for their strong organisational effort and the Scientific and Local Organising Committees for the great care put in the preparation of the event.

We are indebted to Cesare Valenti for the great help given in the editing of the book.

Rosolino Buccheri
Vito Di Gesù
Metod Saniga

Participants

Marco Alfano, *Centre for Communication Studies, Palermo, Italy.*
Richard Amoroso, *Noetic Institute, Orinda, CA, USA.*
Jorg D. Becker, *Universitaet Bundeswehr, Muenchen, Germany.*
Viviana Beninati, *Student, University of Palermo, Italy.*
Serena Billé, *Student, University of Palermo, Italy.*
Richard Block, *Montana State University, Dept. of Psychology, MT, USA.*
Rosolino Buccheri, *CNR, Istituto di Fisica Cosmica, Palermo, Italy.*
Barbara Caci, *Student, University of Palermo, Italy.*
Maurizio Cardaci, *University of Palermo, Dept. of Psychology, Italy.*
Antonella Chifari, *CNR, Istituto di Tecnologie Didattiche e Formative, Palermo, Italy.*
Antonella D'Amico, *CNR, Istituto di Tecnologie Didattiche e Formative, Palermo, Italy.*
Vito Di Gesù, *University of Palermo, Dept. of Mathematics, Italy.*
Benedetto Di Paola, *University of Palermo, Dept. of Mathematics, Italy.*
Vera Maura Fernandez De Lima, *Universidad Campinas, Brasil.*
Stefano Ficata, *Physician, Palermo, Italy.*
Sergio Folchitto, *AGIP Petroli, Roma, Italy.*
Giuseppa Gambino, *Student, University of Palermo, Italy.*
Ornella Gervasi, *University of Palermo, Dept. of Psychology, Italy.*
Ronald Gruber, *Stanford University, CA, USA.*
Maria Guarnera, *Student, University of Palermo, Italy.*
Wolfgang Hanke, *Hehenheim University, Dept. of Physiology, Stuttgart, Germany.*
Pietro Iacono Quarantino, *Student, University of Palermo, Italy.*
Zachary Jacobson, *Carleton University, Ottawa, Canada.*

George Jaroszkiewicz, *University of Nottingham, Dept. of Mathematics, U.K.*
Monica Mandalà, *Student, University of Palermo, Italy.*
Gloria Messina, *Student, University of Palermo, Italy.*
Linda Messineo, *Student, University of Palermo, Italy.*
Raffaella Misuraca, *University of Palermo, Dept. of Psychology, Italy.*
Alfredo Nocera, *Aeronautical Engineer, Palermo, Italy.*
Tartara Osca, *Student, University of Palermo, Italy.*
Simona Ottaviano, *CNR, Istituto di Tecnologie Didattiche e Formative, Palermo, Italy.*
Alfredo Pereira, *Universidade Estadual Paulista, Sao Paulo, Brasil.*
Maria Alice Pereira, *Universidade de Sao Paulo, Brasil.*
Gaia Randazzo, *Student, University of Palermo, Italy.*
Gianni Rigamonti, *University of Palermo, Dept. of Philosophy, Italy.*
Metod Saniga, *Slovak Academy of Science, The Slovak Republic.*
Jonathan Smith, *Iowa State University, Dept. of Mathematics, Ames, IA, USA.*
Filippo Spagnolo, *University of Palermo, Dept. of Mathematics, Italy.*
Tecla Sportelli, *University of Palermo, Dept. of Mathematics, Italy.*
Mark Stuckey, *Elizabethtown College, Elizabethtown, PA, USA.*
Domenico Tegolo, *University of Palermo, Dept. of Mathematics, Italy.*
Giuseppe Trautteur, *University of Naples, Dept. of Physical Sciences, Italy.*
Cesare Fabio Valenti, *University of Palermo, Dept. of Mathematics, Italy.*
Richard Walker, *Second University of Naples, Italy.*
Giovanna Zannelli, *University of Palermo, Dept. of Psychology, Italy.*
Danilo Zavrtanik, *Nova Goric Polytechnic, Slovenia.*

Contents

Introduction

1. Biological and Psychological Aspects of Time

INTRODUCTION

Studies on the structure of time

An interdisciplinary approach looking at new scientific investigation methodologies

ROSOLINO BUCCHERI

Istituto di Fisica Cosmica e Applicazioni dell'Informatica, Area di Ricerca di Palermo del Consiglio Nazionale delle Ricerche, Italy

1. PREMISE

We all know about the intriguing discrepancy between our subjective experience of time and how time itself is described in Physics. We know, in particular, about the existence in our perception of a privileged direction of time in which we may consciously assume the control of the present, the only moment of which we have a direct perception. We also know that time in modern physics does not agree with our experience: the laws of physics are generally time reversible not distinguishing between past and future and ignoring the concept of the present itself, circumstance that relegates all human common experience about perception of time to a simple phenomenological datum not worth of scientific discussion. It is also ignored, in physics, the important difference shown by past and future in our mind, i.e. that the former, being considered as fixed and definite, can just be recalled into the memory, while the latter, being viewed as unknown and undetermined, can be predicted only approximately by taking into account our knowledge about cause-effect relationships.

We are left in practice with two different, and conflictual, concepts of time: an "internal" one, proper of each human being but strictly individual and quantitatively incomparable with that of all other human beings, and an "external" one, defined in physics as a fixed succession of equal instants, independent from any human perception, only depending on the presence of gravitational masses and relative velocities of observers, with no difference

Studies on the Structure of Time: From Physics to Psycho(path)ology
Edited by Buccheri *et al.*, Kluwer Academic / Plenum Publishers, New York, 2000.

between past and future, only useful as a parameter in the equations of motion.

Can we reconcile the two concepts? Is there any relation between the "illusive" time of Physics, classically considered of secondary importance in the description of the laws of nature, and the "real" time of Biology, certainly very important for any evolutionary process?

The present workshop aims to approach the question starting from two standpoints. One refers to the psychology of the perception of time (particularly in the altered states of consciousness) using the quite large quantity of literature data made available in the course of the closing century, the other refers to those achievements of modern physics related to the causality breakdowns implied, in some cases, by quantum mechanics.

A lateral but not less important goal of the meeting is to try to set up an international interdisciplinary scientific cooperation aiming to study the comparison between the two concepts of time from different perspectives, all converging to the objective of better understanding the nature of time itself beyond the pseudo-certainties of the current science.

The initiative, undertaken by Metod Saniga and myself in the context of the recent revival of interest in the questions related to time, has collected the consensus of several Institutions to whom I would like to express my sincere gratitude; it is this consensus together with the personal contribution of many scientific personalities, for whom these topics form the subject of their current scientific research, that has made possible this meeting and all the productive discussions which followed.

Myself, I have not done relevant systematic research in this specific field. As an astrophysicist I have been working in high energy Astrophysics for about twenty years, my interests having been focalised in the study of gamma-ray emission from compact stars. Unfortunately (or fortunately) around 1991 I left gradually Astrophysics and moved to an activity of administration and organisation of scientific research with the peculiar task of promoting interaction among different scientific disciplines and between the research community and the external world, especially of the political, economical and academic ones.

It was unfortunate because my new activity soon started to be very demanding in time and effort such that I had to abandon the active research. On the other side, this new situation allowed me to devote more attention to wider scientific objectives, beyond the restricted field in which I had to build my scientific career.

Now that I have completed my task, by releasing the new Area della Ricerca di Palermo where all the CNR Institutes can operate from now on with new and technologically advanced scientific and organisational facilities, I hope to return to be more deeply and directly involved in active research. I hope, as well, that this meeting will generate further scientific developments for the study of the enigma of time, by treasuring all variety of experiences from researchers and thinkers in many disciplines.

2. CLASSICAL PHYSICS, ILLUSIVE TIME AND DETERMINISM

Man has dedicated a lot of intellectual effort to understand the nature of time. The question, from various standpoints and with different formulations, has been the subject of a debate in philosophy (and in physics) for two and a half millennia, having started historically around 450 B.C. with the dispute between the Eraclito's *Pànta Réi* in which "nobody can dip his feet twice in the same river" and the Parmenide's immobile and indestructible *tò eon* whose duration exclude past and future and compresses everything in a pure present. The matter was reconsidered by Plato in *Timeo* and in *Parmenides* and later on by Aristotle in *Physics.*

In Plato's *Timeo* the flow of the time, described as the image of the eternity moving ahead according to numbers, is indicated by the perennial motion of the stars. According to Plato, all the observed events in space and time are a visible projection of our invisible soul, thus establishing a connection between Time and Psyche, connection which will remain a constant characteristic in the western culture.

From then on, a vast literature on the nature of time was originated, based on philosophical, psychological and physical arguments. A detailed study was done by Aristotle in *Physics*, where time is described as "*a number of motion with respect to before and after... which is permeating everything, and which only soul or mind are able to put into sequence*" therefore considering impossible the existence of time without that of soul.

Aristotle's observations and opinions became a reference point of view for centuries ahead and also the Catholic Church, in the following, used Aristotelian models of nature, particularly of time, which was reanalysed from the religious standpoint by Saint Augustine. Until when Galileo and Newton defined the new scientific method, Aristotelian time was considered to have a fundamental role in the evolution of the mankind, marking all the events of the universe from their origins to their ends.

Starting with Galileo and Newton and all along the years of positivism, the important achievements in physics stimulated the opinion (still now deep-rooted in many scientists) that the laws of mechanics could explain all observed phenomena and that, since they are independent from any privileged direction of time, present, past and future as perceived by humans, are pure illusion, somehow created by our mind. Mechanistic determinism became the current way of thinking: past and future of any physical event are perfectly describable at any time, provided that all physical conditions of the event itself are known in details and in the assumption that the observer is viewed *à la Descartès*, as external to nature.

The mechanicistic paradigm set up by Newtonian science found in Pierre Simon de Laplace his clearest expression:

"*Anyone who at a given instant could be aware of all the forces of the nature and the reciprocal situation of everything in it, if furthermore, he could have the ability to submit to detailed analysis all these items, he would embrace in the same formula both the motion of larger bodies of the universe and that of the lighter atom: nothing would be uncertain for him and the future as well as the past would simultaneously be present to his eyes*".

Many other famous and important scientists have been bound to this paradigm and do not recognise any scientific validity to the division of time in past, present and future regions.

Paul Davies thinks that this division has no scientific basis and that space and time "*are destined to disappear like ghosts*";

Richard Feynman was convinced that science will arrive at a point where all the essential laws of nature will be known;

Stephen Hawking predicts the achievement of a unified law which "*will be able to decode God's mind*".

In Astrophysics, the Big Bang theory, with the achievement of an encouraging degree of presumed irrefutable certainties seems to have closed a cycle and the proposed birth of space and time seems to have reached one of the highest ranks of importance in the human discoveries.

Einstein himself condemned the obstinacy to distinguish different regions of time and confirmed the interpretation that time is illusion on the main basis that the laws of Physics do not contemplate irreversibility of physical processes and therefore cannot describe any privileged direction from past to future. It is fair to say, however, that the Einstein position was taken very seriously by Kurt Gödel who proposed a cosmological model in which it was possible to travel back to the past, model to which Einstein replied saying that he could not believe to this possibility and adding that it should stimulate physicists to reconsider the problem of irreversibility.

But more than the new Einstein position, the acknowledgement that the ancient dispute about the nature of time was not at all overcome is coming from the ongoing discoveries about evolution of complex (and living) systems, together with a psychological refusal of the extreme simplifications of determinism and its implications that history, novelty and creativity could not have any room in our life.

3. IRREVERSIBILITY IN PHYSICS, EVOLUTION AND ARROW OF TIME

In Science, the concept of a privileged direction of time (called "arrow of time" after Arthur Stanley Eddington) was born in the framework of thermodynamics after the postulated increase of entropy in isolated systems and therefore of the constant increase of their disorder. Ludwig Boltzman, with his mathematical definition of Entropy, expressed as the probability of formation of a given macroscopic configuration of particles, supplied physics with an effective instrument, able to predict the evolution in time of an isolated system toward equilibrium.

Thermodynamics, however, being mainly related with physical situations at equilibrium, was considered a field of secondary importance and so was the thermodynamical arrow of time. As a matter of fact, systems at equilibrium do not have any history, all occasional fluctuations tending to disappear, leaving always the system in one, well determined, final situation.

An arrow of time, on the other hand, is strictly connected with the concept of evolution which, in turn, is connected with the concept of irreversibility. The evolutionism movement developed at the end of last century, mainly in Biology but more recently also in Geology and Chemistry, was therefore the picklock for pulling up the arrow of time from the phenomenological ghetto where it had been relegated.

A relevant reaction to positivism came from Henri Bergson. According to him, all psychical events live in a qualitative dimension which cannot be related to the quantitative dimension of physical events. The time concretely lived in our consciousness is a real duration in which the present psychical state is intimately connected with the process from which it is derived: all states of consciousness interpenetrate each other, giving rise to an amalgam in continuous evolution. This really lived process cannot be explained by science with its abstracts and rigid concepts. In this sense Bergson agreed with Einstein when saying that "time is out of Physics". Bergson refuses determinism and its related finalist character because both deny the spontaneity and the creativity of real processes. Positivism interpreted time as a mathematical parameter, an abstract succession of static instants, all equals and indifferent to the qualitative nature of the facts contained in them; Bergson, in polemic with Einstein, refused this kind of spatialised time and proposed a description of the states of consciousness through introspection, with the intention to find a biological model for an interpretation of time that escapes science and, in particular, that escapes the experimental psychology of positivism which pretends to relate the data internal to consciousness to external physical events.

With Ilya Prigogine's theory of dissipative structures operating far from equilibrium, the irreversible processes start to play a constructive and unpredictable role; far from equilibrium, in fact, all fluctuations amplify following chaos laws and creating always new and unpredictable situations, mathematically expressed by non linear equations with many solutions and therefore leading to many possible different situations.

Using Fritjof Capra's language, catalytic cycles, indispensable to living organisms, lead to chaotic instabilities by means of repeated retroactions of self-amplification and, in successive bifurcation points, allow the arising of always new structures of increasing complexity. Irreversibility generates order from chaos implying indetermination as one of the fundamental characteristics of natural phenomena.

Prigogine, therefore, carries Bergson's position in a more scientific context for he considers time as a measure of the evolution of irreversible processes. He is sure that time can be object of science provided that the laws of physics are reviewed with an efficient integration of gravitation and thermodynamics. Prigogine thinks, in particular, that the true problem consists in a correct formulation of the laws of physics, beyond the long time sought unification affected by the deterministic point of view. It is not true that time is out of Physics, it is Physics that is not been able to incorporate time. Time is real but is not well represented in Physics. Prigogine thinks in addition that all real processes are irreversible in nature (demonstrated by the observed evolution in cosmology, in biology, in thermodynamics and in mechanics when passive resistances are taken into account) and that this irreversibility is what can give to us the Aristotle's perspective of the "before and after".

In agreement with Saint Augustine, time is therefore, for Prigogine, certainly antecedent to the birth of the universe. In the Prigogine's paradigm, all evolves in time and increases in complexity: more complex the system, more unpredictable its future. Our universe should have been born from a process of self-organisation, set up at the first stage of the appearance of a chaotic instability of the pre-universe, induced by the interaction between gravitation and matter. The big bang, as the pre-eminent irreversible process, cannot be reproducible and cannot be the start of the space-time but just the passage from one situation to another, unpredictable, as we have to expect from such a system. In other words the universe, being the most complex existing system, is also the most unpredictable one and therefore any theory predicting the future of the universe is superficial, at least with the present knowledge. We need to incorporate irreversibility of macroscopic phenomena in the laws of nature and give the correct importance to the arrow of time.

It is not useless to point out here, that a negative role in decreasing the importance of the arrow of time could have been played by the fact that the derivation of the laws of Physics had always been based on an efficient schematisation of real situations, the so called "ideal" experiments in which energy dissipation processes like friction or air resistance were neglected, thus excluding the privileged direction of time implicit in the second principle of thermodynamics.

Even more important, in obtaining the reversibility of the parameter "time" in the equations of Physics (base of determinism) could have been the science's attitude to consider the observer independent of the observed world. As a matter of fact, a full inclusion of a complex system like the observer in the description of the laws of nature renders difficult, at least with the present knowledge, any idealisation and related approximations together with the establishment of precise equations of motion. By including the observer, the evolution of nature becomes practically impossible to predict, and the future becomes formally different from the past.

The concept that the measure of some physical parameters (like duration and lengths) might depend on the observer was first included in the Einstein's relativity and soon after reinforced by the uncertainty principle in quantum mechanics where also the modalities of observations became important. In Fritjof Capra's ecological theories also the consciousness of the observer seems to acquire importance.

John Wheeler (in accordance with Boltzman) said that the observer and his consciousness create time and that the latter would not exist in an universe without observers. For Boltzman who, for years, had desperately tried to understand, without success, the reason of irreversibility of macroscopic phenomena, a solution of the problem was the existence in the universe of thinking beings capable to perceive in their unconscious the distinction between past and future. In this way Boltzman and Wheeler supported the hypothesis that the characteristics of our universe (whatever its origin) are related to the existence of a conscious observer, thus anticipating the concept of "antropic principle" later formulated by Brandon Carter.

Prigogine reverses the concept; according to him, time is an essential constituent of the universe in which the complexity tends to increase according to the laws of chaos but in a creative way, and the human beings (as well as all other irreversible processes) are a product of time of which consciousness takes note of. We must therefore find in science the fundamental role that time has. Reversibility is a simplification that limits the development of science.

unpredictable but still deterministic solutions? in this last case, the William James paradox of determinism would remain unsolved and we could still not be able to decide whether every event is more or less rigidly caused by another preceding one or, in contrast to it, all of us have always the possibility to choose freely among many possible alternatives.

– finally, going into details of the human internal time, how can we reconcile the opposite experiences of those living the altered states of consciousness? In other words, how is it possible that individual perceptions of time duration are so different from each other and from an objective datum? how can be explained, as an example, the a-temporality of mysticians or the absence of past or future described by mentally sick people?

Any complete description of the physical world cannot escape unambiguous answers to these questions. The workshop intends to give a contribute trying to clarify some terms of the problem (see for example the discussion on the biological and psychological aspects of the perception of time - first session, the experiment at CERN about the first direct observation of time-reversal non-invariance and the discussion about bergsonian time, chaos, entropy - second session) and, perhaps, suggesting some future lines of action where to address future research. The scientific use of literature data concerning "altered states of consciousness" for modelling the flow and the characteristics of the perceived "internal" time, or the studies on pre-geometry and on discrete space-time, may constitute a new way for the establishment of more effective methodologies of scientific investigation, aiming to give new insight for the solution of the enigma of time.

REFERENCES

1. St. Augustine, 402, *Confessioni.*
2. Bergson H., 1889, *Saggio sui dati immediati della coscienza.*
3. Borges J.L., 1963, *Otras Inquisiciones.*
4. Capra F., 1996, *The web of life.*
5. Davies P., 1995, *About time - Einstein unfinished revolution.*
6. Eddington A.S., 1927, *The nature of the physical world.*
7. Feynman R., 1965, *The character of Physical Law*, MIT.
8. Fisher E.L., 1995, *Aristotles, Einstein & Co.*
9. Hawking S., 1988, *From Big Bang to Black Holes*, Bantam.
10. James W., 1956, *The Will to believe.*
11. Laplace P.S. de, 1967, *Opere*, UTET.
12. Prigogine I., 1996, *La fin des certitudes. Temps, Chaos et les lois de la Nature.*
13. Smolin L., 1997, *The life of the Cosmos.*

1.

BIOLOGICAL AND PSYCHOLOGICAL ASPECTS OF TIME

Psychological Time

Memory, Attention, and Consciousness

[1]RICHARD A. BLOCK AND [2]DAN ZAKAY
[1]*Montana State University, Bozeman, Montana, U.S.A.*
[2]*Tel Aviv University, Tel Aviv, Israel*

1. INTRODUCTION

All animals must be able to encode temporal properties of events, store representations of those properties, and use those representations for actions. Although no sensory organ or perceptual system uniquely subserves psychological time[1], temporal properties of events provide important information[2,3]. More highly evolved animals can also consciously remember the timing of past events and anticipate the timing of future events. They experience several aspects of time, including temporal location, duration, and temporal perspective[4,5]. Timing of events lasting from seconds to minutes is especially important for survival. Many psychologists have studied duration timing, probably because it is a complex and important aspect in terms of environmental adaptation. Most theorists explain duration timing in terms of cognitive processes, biological processes, or both. Some aspects of duration timing require attention and memory, and the study of duration judgments reveals and clarifies the underlying processes[6,7].

In the first section, we review cognitive evidence on various kinds of temporal judgments and show how they are interrelated. In the second section, we review physiological and neuropsychological evidence on brain processes that subserve psychological time.

Studies on the Structure of Time: From Physics to Psycho(path)ology
Edited by Buccheri *et al.*, Kluwer Academic / Plenum Publishers, New York, 2000.

2. COGNITIVE EVIDENCE

2.1 Temporal Location Judgments

Researchers who have studied temporal location judgments have made some relevant discoveries. In an early experiment[8], participants were told to pay attention to a series of words for a later memory test. They were not warned that temporal judgments would be required. After all the words had been presented, participants were asked to judge the temporal position of each word in the series. As shown in Figure 1, if they remembered the word, they were able to judge its temporal position with some accuracy. Temporal information is encoded in memory even under incidental conditions, and this process apparently does not require the conscious intention to encode temporal information.

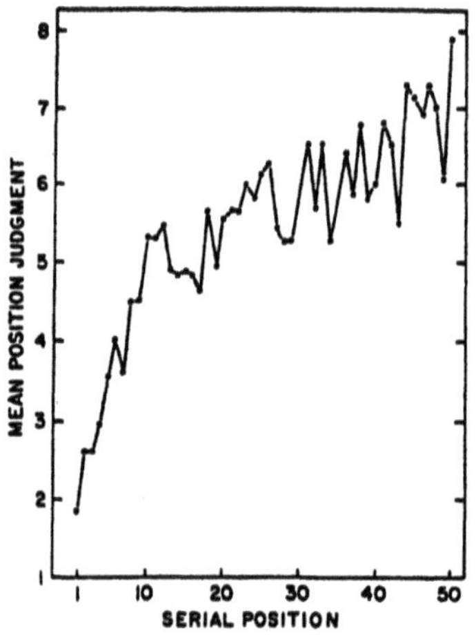

Figure 1. Mean judgment of temporal position judgment (on a scale from 1 to 10) as a function of the serial position of a word in a series of words. From "Repetition and memory: Evidence for a multiple-trace hypothesis", by Hintzman and Block, 1971, Journal of Experimental Psychology, 88, p. 299. Copyright 1971 by the American Psychological Association. Reprinted with permission.

The findings of several subsequent experiments clarified this finding. In one of them[9], two separate series of words were presented, again under incidental conditions. Afterwards, the participants judged whether each word had occurred in the first series of words or the second, and then whether it had occurred near the beginning, middle, or end of the series. As before, they were able to make these judgments with considerable accuracy. The kinds of errors that they made were revealing, however. When the participants incorrectly judged that a word had occurred in a particular series, they nevertheless tended to judge that it had occurred in the correct part of the series. Thus, remembered events did not simply migrate to temporally adjacent locations. This finding suggests that temporal position judgments are based on incidentally encoded contextual information instead of on hypothetical time tags. As such, contextual information enables people to locate events on a relative scale of psychological time, not on a continuous scale of absolute time. Other studies provided additional evidence that people use contextual information to make temporal position judgments. In one, each word was presented either auditorily or visually. Although the participants were not forewarned, they were able to remember the presentation modality of words with above-chance accuracy[10]. They could even remember whether an auditory presentation of a word had occurred earlier than a visual presentation, or vice-versa[11].

Converging evidence comes from studies of autobiographical memory, in which researchers asked participants to date personal memories of events that occurred relatively long ago, such as months or years in the past. People can temporally locate such personal memories with some accuracy. However, their judgments also reveal systematic errors, which are called *scale effects*: People may accurately remember an event as having occurred during a particular time of day but inaccurately remember the day, month, or year during which the event occurred[12]. This evidence supports the view that temporal memory judgments involve location-based rather than distance-based processes[13]. Location-based processes involve judging the recency of an event in a way that is influenced by important contextual landmarks, whereas distance-based processes are those that involve judging recency in a more absolute way, such as in terms of absolute time tags. Evidence for location-based processes suggests that relative contextual information, rather than absolute time tags, are used to make temporal memory judgments[14].

Contextual information is apparently encoded in memory relatively automatically, without requiring a person's conscious intent to do so. The evidence comes from two main sources. First, people can reasonably accurately judge the temporal location of an event although they were not warned that they would have to do so[8]. Second, some researchers have compared incidental and intentional conditions. Under intentional conditions, participants are forewarned that they will be asked to remember the temporal location of events. These studies show that there is sometimes not much greater accuracy under intentional conditions[15]. When people remember the temporal location of past events, they rely on contextual information and logical inferences based on it.

Memory for the temporal location of past events is apparently not limited to humans. Birds such as scrub jays are able to remember how long ago they cached certain food items--that is, how long ago they stored them for later use[16,17]. This evidence meets the criteria for episodic-like memory in non-human animals--that is, memory for contextual associations to personally experienced events, which in this case was the act of caching the food items. Remembering the temporal location of past events involves "mental time travel," and the underlying processes are not limited to humans.

2.2 Duration Judgments

Two kinds of duration judgments must be distinguished: judging the duration of a single event, and judging the duration of an entire series of events. These two kinds of duration judgment almost certainly involve different processes.

Animals such as pigeons and rats can learn to make one response to a relatively short-duration stimulus and another response to a relatively long-duration stimulus[18]. People can remember the approximate duration of each event in a long series of events even if they are not forewarned about the task[19]. Duration information is apparently encoded relatively automatically as an integral part of the experience of an event.

More researchers, however, have investigated the processes involved in judging the duration of a series of events. In this case, duration is not simply an integral property of a single event, and the processes involved in experiencing and remembering the duration of a series of events are probably consciously controlled. Psychological time depends on complex interactions among conditions prevailing when a duration is experienced and when it is judged[20]. James made an important distinction in his well-known chapter on time perception. He claimed that different variables influence the "retrospective and prospective sense of time"[21]. Researchers have recently investigated this claim, using two kinds of methodologies. In one kind, participants prospectively expose one stimulus per second until stopped by the experimenter, and then they retrospectively estimate the total duration. The duration produced prospectively is usually greater than the retrospective verbal (numerical) estimate[22,23]. Because participants judge the duration with two different methods, this methodology does not afford a valid comparison of prospective and retrospective timing.

Other researchers have studied James's distinction by comparing duration judgments in what are now called the *prospective paradigm* and the *retrospective paradigm*. The prospective paradigm involves experienced duration, whereas the retrospective paradigm involves remembered duration[5]. In the former, participants have advance knowledge that they will be asked to make a duration judgment; in the latter, they do not. In both, participants experience a duration containing the same external information and processing task. In the prospective paradigm, people may intentionally encode temporal information along with information about the contents of the duration. In the retrospective paradigm, people may automatically encode contextual information along with information about the contents of the duration; later, they must retrieve information from memory to make a duration judgment.

Many researchers have used the prospective paradigm, but few have used the retrospective paradigm. The main reason is that after a person is asked to provide a single retrospective judgment the paradigm becomes prospective. (Investigating the retrospective paradigm prohibits obtaining repeated measures, and comparing the two paradigms requires a between-subjects design.) In an early, influential study[24], participants sorted ordinary playing cards according to various rules that required the processing of 1, 2, or 3 bits of information per card. In the retrospective paradigm, the amount of information processed did not influence duration judgments. However, in the prospective paradigm, the amount of information processed was inversely related to subsequent duration judgments. In the following sections, we discuss the interpretations of these findings in more detail.

Subsequent experiments revealed that prospective judgments tend to be larger in magnitude and smaller in variability than retrospective judgments[25]. More importantly, several variables influence duration judgments differently in the two paradigms. For example, experienced duration shortens if a person had performed a more difficult processing task, but remembered duration is not affected[26]. In addition, remembered duration lengthens if a person had performed several different kinds of tasks, but experienced duration is not affected. These findings provide strong evidence that different processes or systems subserve duration judgments in the two paradigms. Under retrospective conditions, people apparently judge a duration based on the contextual changes that were automatically encoded in memory[27,28]. Under prospective conditions, contextual changes are also automatically encoded, of course, but they apparently play a minor role. The reason is that under prospective conditions participants consciously attend to time and thereby accumulate salient temporal information.

2.3 Retrospective Duration Judgments: Remembered Duration

In the retrospective paradigm, people probably do not attend to time very much unless there is little information to process, a boring situation, or similar conditions. Consequently, most theorists have proposed memory-based models of remembered duration. Specific theorists have focused on the "multitudinousness of the memories which the time affords"[21]; stored and retrieved information, or "storage size"[29]; remembered changes[30]; encoded and retrieved contextual changes[28], or interval segmentation[31]. People may judge a duration as being longer in retrospect if they have processed and can remember a greater number of stimuli or more complex stimuli[29]. However, remembered duration is not based simply on the recallability of individual events or their encoded complexity[28,32]. People probably do not attempt to retrieve memories of all events from the duration. Instead, they remember a duration as being longer to the extent that they can easily retrieve some events, relying on an availability heuristic.

Memory for changes in cognitive context has an important influence on remembered duration. Contextual changes occur as a result of variation in environment stimuli, internal stimuli, information processing strategies, and other such factors[20]. For example, people remember a duration as being longer to the extent that there were greater changes in processing strategies used to encode information. Environmental context is another salient source of contextual changes: Previous experience in a particular environment shortens the remembered duration of a subsequent time period spent in it[27]. Remembered duration is longer if a duration is segmented by high-priority events which attract attention, such as politicians' names inserted among names of animals[33]. Segmentation, however, may simply cause contextual changes that result from the appearance of the high-priority events. These findings lead to the rejection of event-memory explanations of remembered duration, such as a storage-size model[29].

Retrospective duration judgments typically show what is called a *positive time-order effect*, in which the first of two equal time periods is remembered as being longer than the second[27,34]. This effect is somewhat counterintuitive: A storage-size model predicts the opposite, a negative time-order effect attributable to items dropping out of storage[29]. However, a greater number of changes in contextual elements may be encoded during a more novel experience, such as during the first duration. Several findings support the notion that contextual changes underlie the positive time-order effect. The effect is eliminated if the environmental context prevailing during the second duration is different from that prevailing during the first[27]. It is also eliminated if changes in emotional context that would ordinarily occur during the first duration occur instead during a preceding time period[35]. If the interval between two durations increases, so that the first one becomes relatively less recent and the second one becomes relatively more recent, people may have more difficulty remembering the contextual changes that occurred during the first duration. In this case, the typical positive time-order effect may reverse, becoming a negative time-order effect[36].

2.4 Prospective Duration Judgments: Experienced Duration

Prospective duration timing is sensitive to attention-demanding processes that occur concurrently with the processing of nontemporal (task) information. Experienced duration lengthens if there are fewer stimuli requiring processing or if a processing task is easy[7,24,37]. Prospective timing is essentially a dual-task condition in which attention is shared between temporal and nontemporal information processing. Attending to time requires some of the same attentional resources that attending to nontemporal (stimulus) information does. For this reason, most theorists emphasize the role of attentional resource allocation[6,38,39,40,41]: Experienced duration increases to the extent that a person is able to allocate relatively more resources to processing temporal information.

Evidence from research using a dual-task paradigm supports an attentional model. If a person is told how much attention to allocate for stimulus information processing and how much to allocate for temporal information processing, prospective duration judgments are a function of the latter[39,42]. The magnitude of prospective duration judgments increases if participants are told that a temporal task is the primary task and that a simultaneous nontemporal task is the secondary task[43]. Duration judgments decrease when the instructions are reversed. If a nontemporal stimulus that attracts attention is presented, prospective duration judgments decrease[44]. In addition, if participants are asked to monitor the duration of several concurrent events, timing accuracy decreases as a function of the number of monitored events[42].

When people make prospective duration judgments, they compare accumulated temporal units with the typical number of temporal units stored during reference durations. Figure 2 shows one model of this, the attentional-gate mode[17,41]. In it, a pacemaker produces pulses. When one attends to time instead of to external events, the pulse stream is sent to subsequent components. A cognitive counter accumulates a total pulse count during the duration, and this total number of temporal units is compared to relevant total counts stored in reference memory. Reference memory is a long-term store containing learned temporal units and translations of these total temporal units into verbal (numerical) units. Some of these total counts may originally have been stored in a retrospective way. For example, people who repeatedly encounter a specific traffic (stop) light may initially encode relevant duration retrospectively. Later, they may attend to time in a prospective way, storing the total temporal units for this particular duration. As people acquire expertise with a duration, their duration estimates presumably become more accurate and less variable[45].

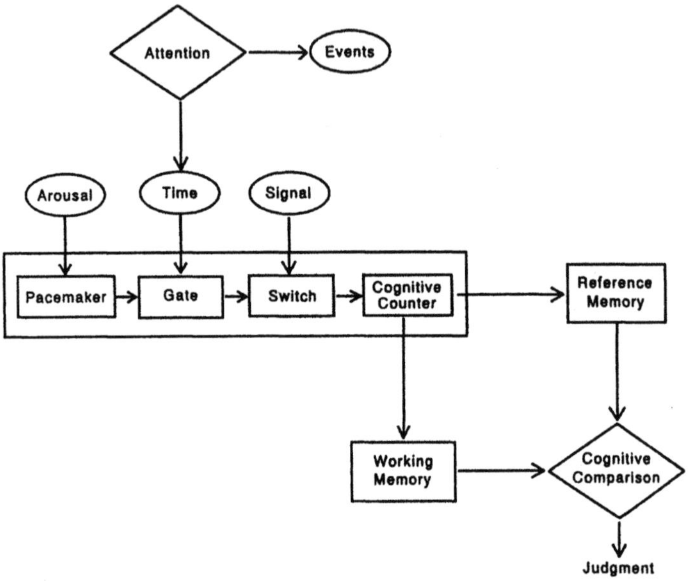

Figure 2. Attentional-gate model of prospective duration judgment. Adapted from "Temporal cognition", by Zakay and Block, 1997, Current Directions in Psychological Science, 6, p. 14. Copyright 1977 by the Cambridge University Press. Adapted with permission.

3. PHYSIOLOGICAL AND NEUROPSYCHOLOGICAL EVIDENCE

The physiological processes that underlie temporal experiences are beginning to be revealed. Converging evidence comes from case studies of neuropsychological patients, new methods of brain scanning, and experiments on the effects of psychoactive drugs[46]. We first discuss processes underlying temporal location judgments, then those underlying remembered duration, and finally those underlying experienced duration.

3.1 Temporal Location Judgments

Patients with damage in the frontal lobes of the brain usually show relatively little impairment in remembering events, but they usually show severe impairment in remembering temporal information. For example, they often cannot remember which of two events occurred more recently[47,48,49].

This impairment occurs mainly if there is damage to the dorsolateral prefrontal cortex in and around Brodmann's area 46 (see Figure 3). There are at least two possible explanations[49]. First, the dorsolateral prefrontal cortex may directly encode contextual information that allows a person to make inferences about temporal order of events. Second, the dorsolateral prefrontal cortex may mainly be involved in organizing conscious processes required for retrieval of information from memory[50,51]. This function may be somewhat different from that involved in generating contextual information, although evidence on that issue is unclear. These two functions are not necessarily mutually exclusive: The prefrontal cortex may be critically involved in both the encoding and the retrieval of contextual information. It is clear also that the prefrontal cortex is especially involved in dynamic control of behavior across time[52,53]. As such, it is an integral part of the *working memory* or *working with memory* system of the human brain[54], a system that is intimately linked to consciousness.

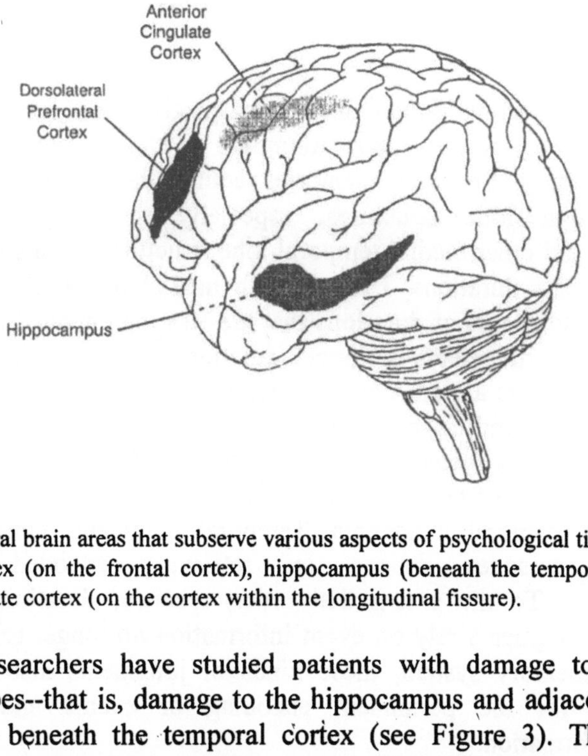

Figure 3. Several brain areas that subserve various aspects of psychological time: dorsolateral prefrontal cortex (on the frontal cortex), hippocampus (beneath the temporal cortex), and anterior cingulate cortex (on the cortex within the longitudinal fissure).

Some researchers have studied patients with damage to the medial temporal lobes--that is, damage to the hippocampus and adjacent structures located just beneath the temporal cortex (see Figure 3). These patients experienced a series of events. Then test stimuli were presented in pairs, and the patients had to judge which of the two stimuli had occurred more recently. Although these patients were often unable to remember both stimuli, when they did remember both of them they performed normally on the recency judgment task[55]. Thus, although patients with medial temporal-lobe damage show deficits in encoding permanent memories, they often can remember associated contextual information, which they may use to make temporal judgments about the events. As we mentioned earlier, frontal-lobe patients show roughly the opposite: impaired memory for temporal information but normal memory for event information. This suggests that the hippocampus and the dorsolateral prefrontal cortex may perform separate but interrelated functions. Specifically, it suggests that "remembering of temporal order constitutes one component of episodic memory, subserved by the frontal regions, and that remembering of recently presented items constitutes another component of episodic memory, likely subserved by the medial temporal regions"[56]. The dorsolateral prefrontal cortex may supply the hippocampus with contextual information, and the hippocampus may encode this information in association with information about events per se.

3.2 Retrospective Duration Judgments: Remembered Duration

The physiological underpinnings of retrospective duration judgments are also beginning to be revealed. Abundant evidence suggests that the long-term encoding of episodic memories requires the normal functioning of medial temporal lobe structures, especially the hippocampus. The hippocampus and other medial temporal lobe structures are also involved in remembering past durations. Patients with medial temporal lobe damage make normal judgments of durations that are less than about 15 s but make abnormally short judgments of longer durations[57,58]. The prefrontal cortex and the hippocampus are anatomically connected, and both are involved in the processing of working-memory and episodic-memory information[59,60]. The working-memory system of the dorsolateral prefrontal cortex apparently generates encodings of contextual information, which a person may use to make various temporal judgments. Temporal judgments concerning past events and episodes depend on retrieving encoded contextual changes[26,27,28,35]. To the extent that order, recency, duration, and other retrospective judgments rely on event information no longer represented in the working-memory system, these kinds of judgments also require the hippocampus for the permanent encoding of events and associated contextual information.

3.3 Prospective Duration Judgments: Experienced Duration

Although the dorsolateral prefrontal cortex is intimately involved in psychological time, its exact role remains somewhat unclear[61,62]. Nevertheless, the working-memory system of the dorsolateral prefrontal cortex apparently plays an important role in the conscious experience of time in passing. The dorsolateral prefrontal cortex may contain neural circuits that are part of an internal-clock mechanism that is necessary for experiencing and judging durations in the range of seconds[6,46,63,64].

The neurotransmitter dopamine is an important one in the prefrontal cortex. Evidence suggests that the dopamine D1 receptor site plays an important role in working memory[65,66]. Some drugs that influence experienced duration may influence D1 dopamine receptor sites in the dorsolateral prefrontal cortex. Dopamine agonists lengthen experienced duration, whereas dopamine antagonists shorten it[23]. Researchers who have administered drugs to animals trained on timing schedules have also suggested that dopaminergic neurons may underlie the internal clock. Older brain areas such as the basal ganglia may also be part of the internal-clock mechanism[67].

Possibly related to these observations, some neuropsychological patients who had received bilateral lesions of the dorsomedial nucleus of the thalamus displayed an impairment of temporal orientation called *chronotaraxis*. Chronotaraxis may appear as an inability to know the season, the date, the day of week, or the time of day. Many thalamotomy patients display a more or less pronounced chronotaraxis[68,69,70,71]. The prefrontal cortex is the primary projection area of the dorsomedial nucleus of the thalamus, and the prefrontal cortex projects back to thalamic nuclei. In one way or another, all temporal experience may be mediated by the dorsomedial nucleus of the thalamus[54,72].

The older, mostly German, literature contains several case studies of what has come to be called the *Zeitrafferphänomen*[74], or *accelerated time phenomenon*. A *Zeitraffer* is a simple mechanical apparatus to accelerate the apparent motion in a film, as in time-lapse cinematography. One recent case study focused on a Zeitraffer patient, called BW, with damage in his left-hemisphere prefrontal cortex resulting from a glioblastoma[75] (see Figure 4). BW's symptoms arose suddenly: As he was driving his automobile, he noticed that external objects seemed to be rushing toward him at a very fast rate. (This was much faster than the usual rate at which automobiles seem to move in large cities.) He wisely pulled his car to the side of the road. BW later described his experience as that of an "accelerated motion" of events. He also complained that he could not even watch television because the progression of events was too fast for him to follow. When asked to produce a 60-s duration, BW's production was greatly lengthened (averaging 286 s). This evidence converges with his description of his condition. He experienced events in much the way a normal person would experience events on a videotape being played on fast forward (i.e., occurring at approximately six or seven times their normal rate). The most likely explanation is that the pacemaker component of his internal clock was now producing pulses at a considerably decreased rate. If this is the case, the prefrontal cortex apparently contains neural circuits that are at least partially involved in an internal clock.

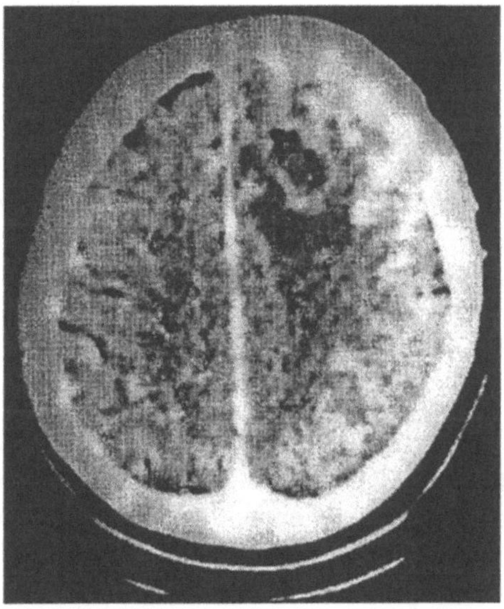

Figure 4. A computerized tomography slice showing the lesion (the relatively dark area) in BW's left-hemisphere prefrontal cortex. Note that (according to standard radiological convention) the left hemisphere is at the right, and the anterior part of the cortex is at the top. (Imagine looking at the brain from below.) Adapted from "Accelerated Time After Left Frontal Cortex Lesion", by Binkofski and Block, 1996, Neurocase, 2, p. 487. Copyright 1996 by the Oxford University Press. Adapted with permission.

Brain areas that subserve a possible internal clock may interconnect with other areas that subserve the attentional processes that underlie experienced duration. However, it is unclear which areas of the brain areas may underlie attention to time. Various aspects of attentionally based task performance activate several brain areas, including the thalamus, the parietal lobes, and the anterior cingulate cortex[76]. These areas probably play somewhat different roles. The most likely candidate for a brain area that subserves the allocation of attention to external events or to time is the anterior cingulate cortex. This brain area is located in the anterior part of the longitudinal fissure dividing the two cerebral hemispheres (see Figure 3). The anterior cingulate cortex may be the essential component of an executive attention network that directly influences the working-memory functions of the dorsolateral prefrontal cortex. By playing this role, it may be the aspect of attention that is involved in the conscious experiencing of durations. Additional research is needed to clarify the roles of the anterior cingulate cortex, the dorsolateral prefrontal cortex, and the hippocampus in various kinds of tasks requiring temporal judgments[77].

4. SUMMARY AND CONCLUSIONS

Psychological time involves the awareness of and the ability to judge several kinds of time-related information. A distinction between various temporal judgments is partly based on whether or not a person is aware that a judgment will be needed. People can make some temporal judgments retrospectively. These judgments concern temporal location (recency, order, and even recognition memory) as well as duration. The relevant information, which is contextual in nature, is encoded relatively automatically. This is revealed by studies in which participants do not know that temporal information will be needed until later (incidental memory). People make other temporal judgments prospectively. These involve situations in which a person is aware that temporal information will be needed, either immediately or later. Although the same kinds of contextual information may be encoded as in a retrospective paradigm, other information is also encoded. This information is only encoded in a deliberate (controlled) way, involving attention and hence more complete involvement of mechanisms of consciousness. Attending to this temporal information apparently competes for the same pool of attentional resources as does attending to nontemporal information.

Evolutionarily newer cortical areas (e.g., dorsolateral prefrontal cortex, hippocampus, and associated structures) subserve the relatively automatic encoding of contextual information and, hence, retrospective temporal judgments. Evidence suggests that the additional involvement of evolutionarily older subcortical (and some newer cortical) structures subserves prospective judgments. Damage to cortical or subcortical brain areas that are involved in psychological time may alter a person's duration experience or temporal perspective.

ACKNOWLEDGEMENT

We thank Ronald Gruber for his comments on this article.

REFERENCES

1. Gibson J.J., 1975, Events are perceivable but time is not, *The study of time II*, Fraser and Lawrence (eds.), Springer-Verlag, New York, 295-301.
2. Klemmer E.T., 1957, Simple reaction time as a function of time uncertainty, *Journal of Experimental Psychology*, **54** 195-200.
3. Michon J.A. and Jackson J.L., 1985, Introduction: The psychology of time, *Time, mind, and behavior*, Michon and Jackson (eds.), Springer-Verlag, Berlin, 2-17.
4. Block R.A., 1979, Time and consciousness, *Aspects of consciousness: Vol. 1. Psychological issues*, Underwood and Stevens (eds.), Academic Press, London, 179-217.
5. Block R.A., 1990, Models of psychological time, *Cognitive models of psychological time*, Block (ed.), Hillsdale, NJ: Erlbaum, 1-35.
6. Block R.A. and Zakay D., 1996, Models of psychological time revisited, *Time and mind*, Helfrich (ed.), WA: Hogrefe and Huber, Kirkland, 171-195.
7. Zakay D. and Block R.A., 1997, Temporal cognition, *Current Directions in Psychological Science*, **6** 12-16.
8. Hintzman D.L. and Block R.A., 1971, Repetition and memory: Evidence for a multiple-trace hypothesis, *Journal of Experimental Psychology*, **88** 297-306.
9. Hintzman D.L., Block R.A., and Summers J.J., 1973, Contextual associations and memory for serial position, *Journal of Experimental Psychology*, **97** 220-229.
10. Hintzman D.L., Block R.A., and Inskeep N.R., 1972, Memory for mode of input, *Journal of Verbal Learning and Verbal Behavior*, **11** 741-749.
11. Hintzman D.L., Block R.A., and Summers J.J., 1973, Modality tags and memory for repetitions: Locus of the spacing effect, *Journal of Verbal Learning and Verbal Behavior*, **12** 229-238.
12. Friedman W.J. and Wilkins A.J., 1985, Scale effects in memory for the time of events, *Memory & Cognition*, **13** 168-175.
13. Friedman W.J., 1993, Memory for the time of past events, *Psychological Bulletin*, **113** 44-66.
14. Shum M.S., 1998, The role of temporal landmarks in autobiographical memory processes, *Psychological Bulletin*, **124** 423-442.
15. Auday B.C., Sullivan C., and Cross H.A., 1988, The effects of constrained rehearsal on judgments of temporal order, *Bulletin of the Psychonomic Society*, **26** 548-551.
16. Clayton N.S. and Dickinson A., 1998, Episodic-like memory during cache recovery by scrub jays, *Nature*, **395** 272-274.
17. Clayton N.S. and Dickinson A., 1999, Memory for the content of caches by scrub jays (*Aphelocoma coerulescens*), *Journal of Experimental Psychology: Animal Behavior Processes*, **25** 82-91.
18. Fetterman J.G., 1995, The psychophysics of remembered duration, *Animal Learning & Behavior*, **23** 49-62.
19. Hintzman D.L., 1970, Effects of repetition and exposure duration on memory, *Journal of Experimental Psychology*, **83** 435-444.

20. Block R.A., 1989, Experiencing and remembering time: Affordances, context, and cognition, *Time and human cognition: A life-span perspective*, Levin and Zakay (eds.), North-Holland, Amsterdam, 333-363.
21. James W., 1890, *The principles of psychology*, Holt, New York, 1 624.
22. Frankenhaeuser M., 1959, *Estimation of time: An experimental study*, Almqvist and Wiksell, Stockholm.
23. Hicks R.E., 1992, Prospective and retrospective judgments of time: A neurobehavioral analysis, *Time, action and cognition: Towards bridging the gap*, Macar, Pouthas, and Friedman (eds.), Netherlands: Kluwer Academic, Dordrecht, 97-108.
24. Hicks R.E., Miller G.W., and Kinsbourne M., 1976, Prospective and retrospective judgments of time as a function of amount of information processed, *American Journal of Psychology*, **89** 719-730.
25. Block R.A. and Zakay D., 1997, Prospective and retrospective duration judgments: A meta-analytic review, *Psychonomic Bulletin & Review*, **4** 184-197.
26. Block R.A., 1992, Prospective and retrospective duration judgment: The role of information processing and memory, *Time, action and cognition: Towards bridging the gap*, Macar, Pouthas, and Friedman (eds.), Netherlands: Kluwer Academic, Dordrecht, 141-152.
27. Block R.A., 1982, Temporal judgments and contextual change, *Journal of Experimental Psychology: Learning, Memory, and Cognition*, **8** 530-544.
28. Block R.A. and Reed M.A., 1978, Remembered duration: Evidence for a contextual-change hypothesis, *Journal of Experimental Psychology: Human Learning and Memory*, **4** 656-665.
29. Ornstein R.E., 1969, *On the experience of time*, Harmondsworth, Penguin, England.
30. Fraisse P., 1963 (Original work published 1957), *The psychology of time* (Leith, Trans.), Harper and Row, NY.
31. Poynter W.D., 1983, Duration judgment and the segmentation of experience, *Memory & Cognition*, **11** 77-82.
32. Block R.A., 1974, Memory and the experience of duration in retrospect, *Memory & Cognition*, **2** 153-160.
33. Poynter W.D., 1989, Judging the duration of time intervals: A process of remembering segments of experience, *Time and human cognition: A life-span perspective*, Levin and Zakay (eds.), North-Holland, Amsterdam, 305-331.
34. Block R.A., 1985, Contextual coding in memory: Studies of remembered duration, *Time, mind, and behavior*, Michon and Jackson (eds.), Springer-Verlag, Berlin, 169-178.
35. Block R.A., 1986, Remembered duration: Imagery processes and contextual encoding, *Acta Psychologica*, **62** 103-122.
36. Wearden J.H. and Ferrara A., 1993, Subjective shortening in humans' memory for stimulus duration, *Quarterly Journal of Experimental Psychology: Comparative and Physiological Psychology*, **46B** 163-186.
37. Zakay D., 1993, The roles of non-temporal information processing load and temporal expectations in children's prospective time estimation, *Acta Psychologica*, **84** 271-280.
38. Brown S.W., 1998, Automaticity versus timesharing in timing and tracking dual-task performance, *Psychological Research*, **61** 71-81.
39. Macar F., Grondin S., and Casini L., 1994, Controlled attention sharing influences time estimation, *Memory & Cognition*, **22** 673-686.
40. Thomas E.A.C. and Weaver W.B., 1975, Cognitive processing and time perception, *Perception & Psychophysics*, **17** 363-367.

41. Zakay D. and Block R.A., 1996, The role of attention in time estimation processes, *Time, internal clocks and movement*, Pastor and Artieda (eds.), North-Holland/Elsevier Science, Amsterdam, 143-164.
42. Brown S.W., 1997, Attentional resources in timing: Interference effects in concurrent temporal and nontemporal working memory tasks, *Perception & Psychophysics*, **59** 1118-1140.
43. Zakay D., 1998, Attention allocation policy influences prospective timing, *Psychonomic Bulletin & Review*, **5** 114-118.
44. Zakay D., 1992, The role of attention in children's time perception, *Journal of Experimental Child Psychology*, **54** 355-371.
45. Zakay D. and Block R.A., The negative asynchrony phenomenon: A prospective timing perspective, *Cognitive contributions to the perception of spatial and temporal events*, Aschersleben, Müsseler, and Bachmann (eds.), Elsevier, Amsterdam.
46. Block R.A., 1996, Psychological time and memory systems of the brain, *Dimensions of time and life*, Fraser and Soulsby (eds.), CT: International Universities Press, Madison, 61-76.
47. Milner B., 1982, Some cognitive effects of frontal-lobe lesions in man, *The neuropsychology of cognitive function*, Broadbent and Weiskrantz (eds.), The Royal Society, London, 211-226.
48. Petrides M. and Milner B., 1982, Deficits on subject-ordered tasks after frontal- and temporal-lobe lesions in man, *Neuropsychologia*, **20** 249-262.
49. Milner B., McAndrews M.P., and Leonard G., 1990, Frontal lobes and memory for the temporal order of recent events, *Cold Spring Harbor Symposia on Quantitative Biology*, **55** 987-994.
50. Mangels J.A., Gershberg F.B., Shimamura A.P., and Knight R.T., 1996, Impaired retrieval from remote memory in patients with frontal lobe damage, *Neuropsychology*, **10** 32-41.
51. Moscovitch M. and Melo B., 1997, Strategic retrieval and the frontal lobes: Evidence from confabulation and amnesia, *Neuropsychologia*, **35** 1017-1034.
52. Fuster J.M., 1995, Temporal processing, *Annals of the New York Academy of Sciences*, **769** 173-181.
53. Fuster J.M., 1997, *The prefrontal cortex: Anatomy, physiology, and neuropsychology of the frontal lobe*, (3rd ed.), Lippincott-Raven, Philadelphia.
54. Moscovitch M., 1992, Memory and working-with-memory: A component process model based on modules and central systems, *Journal of Cognitive Neuroscience*, **4** 257-267.
55. McAndrews M.P., and Milner B., 1991, The frontal cortex and memory for temporal order, *Neuropsychologia*, **29** 849-859.
56. Schacter D.L., 1989, Memory, *Foundations of cognitive science*, Posner (ed.), MIT Press, Cambridge.
57. Richards W., 1973, Time reproductions by H.M., *Acta Psychologica*, **37** 279-282.
58. Williams J.M., Medwedeff C.H., and Haban G., 1989, Memory disorder and subjective time estimation, *Journal of Clinical and Experimental Neuropsychology*, **11** 713-723.
59. Goldman-Rakic P.S., Selemon L.D., and Schwartz M.L., 1984, Dual pathways connecting the dorsolateral prefrontal cortex with the hippocampal formation and parahippocampal cortex in the rhesus monkey, *Neuroscience*, **12** 719-743.
60. Olton D.S., 1989, Frontal cortex, timing and memory, *Neuropsychologia*, **27** 121-130.
61. Nichelli P., 1993, The neuropsychology of human temporal information processing, *Handbook of neuropsychology*, Boller and Grafman (eds.), **8** 339-371.
62. Nichelli P., Clark K., Hollnagel C., and Grafman J., 1995, Duration processing after frontal lobe lesions, *Annals of the New York Academy of Sciences*, **769** 183-190.

63. Church R.M., 1989, Theories of timing behavior, *Contemporary learning theories: Instrumental conditioning theory and the impact of biological constraints on learning,* Klein and Mowrer (eds.), NJ: Erlbaum, Hillsdale, 41-71.
64. Rubia K., Overmeyer S., Taylor E., Brammer M., Williams S., Simmons A., Andrew C., and Bullmore E., 1998, Prefrontal involvement in "temporal bridging" and timing movement, *Neuropsychologia,* **36** 1283-1293.
65. Goldman-Rakic P.S., 1992, Working memory and the mind, *Scientific American,* **267** 111-117.
66. Sawaguchi T. and Goldman-Rakic P.S., 1991, D1 dopamine receptors in prefrontal cortex: Involvement in working memory, *Science,* **251** 947-950.
67. Meck W.H., 1996, Neuropharmacology of timing and time perception, *Cognitive Brain Research,* **3** 227-242.
68. Spiegel E.A., Wycis H.T., Orchinik C.W., and Freed H., 1955, The thalamus and temporal orientation, *Science,* **121** 771-772.
69. Spiegel E.A., Wycis H.T., Orchinik C., and Freed H., 1956, Thalamic chronotaraxis, *American Journal of Psychiatry,* **113** 97-105.
70. Spiegel E.A. and Wycis H.T., 1967, Multiple representation of various functions in the human subcortex, *Confinia Neurologica,* **29** 163-167.
71. Spiegel E.A. and Wycis H.T., 1968, Multiplicity of subcortical localization of various functions, *Journal of Nervous and Mental Disease,* **147** 45-48.
72. Markowitsch H.J., 1992, *Intellectual functions of the brain: An historical perspective,* Hogrefe and Huber,Toronto, Canada.
73. Markowitsch H.J., 1995, Cerebral bases of consciousness: A historical view, *Neuropsychologia,* **33** 1181-1192.
74. Payk T.R., 1977, Störungen des Zeiterlebens bei den endogenen Psychosen [Disturbances of time experience in endogenous psychoses], *Schweizer Archiv für Neurologie, Neurochirurgie und Psychiatrie,* **121** 277-285.
75. Binkofski F. and Block R.A., 1996, Accelerated time after left frontal cortex lesion, *Neurocase,* **2** 485-493.
76. Posner M.I. and Raichle M.E., 1994, *Images of mind,* Freeman, San Francisco.
77. Kesner R.P., 1998, Neural mediation of memory for time: Role of the hippocampus and medial prefrontal cortex, *Psychonomic Bulletin & Review,* **5** 585-596.

The mental clock model

Studies on the estimation of time

MAURIZIO CARDACI
Dipartimento di Psicologia, Università di Palermo, Italy

All scholars who have investigated psychological time have stressed its elusive nature. On the one hand, we have a "smooth" or "effortless" experience of the flow of events[1]. On the other hand, in spite of its familiarity, the temporal perception is hard to grasp. According to Gibson[2], "there is no such thing as the perception of time, but only the perception of events and locomotions". In a quite similar way, Michon[1] has suggested that time should not be considered as "a property of the existential mode of things", but as the "conscious experiential product" of the processes that allow human beings to maintain a dynamic, interactive relation with the flow of events in the external world.

This state of affairs could explain why a global notion of psychological time is an over-simplified concept and no single temporal-judgement paradigm, unitary model or method could account for the variety of temporal phenomena.*

Even if a unified approach to psychological time is lacking, I have tried to extract a list of dualistic categories, based on some Zakay's "methodological dilemmas"[3]. In my opinion, these categories tend to recur in the psychological study on time and can be outlined as follows:

* To study the estimation of time, the researchers have provided a number of different methods: the subject can be asked to give a verbal estimation of the target interval duration; to produce a predefined interval of a given length; to reproduce the target interval by means of some operation; to compare two intervals[3]. In regard to the temporal phenomena which have been considered (without mentioning altered states of consciousness), some researchers, interested in psychophysics, have searched for the hypothetical smallest unit of psychological time (the so-called time quantum). Cognitive psychologists have focused on different processes related to time (perception, attention, memory). Developmental as well as clinical and social psychologists have adopted broad approaches to existential perspectives of temporality.

Studies on the Structure of Time: From Physics to Psycho(path)ology
Edited by Buccheri *et al.*, Kluwer Academic / Plenum Publishers, New York, 2000.

1. PERCEIVED VS. REPRESENTED TIME. This dualism is related to the well-known phenomenological distinction between time as an implicit behavioral tuning of our actions to the dynamics of the surrounding world (this kind of temporality is cognitively impenetrable) and time as an explicit diachronic past-present-future dimension. Related dualistic categories are *experienced duration* vs. *remembered duration*, *simultaneity* vs. *successiveness*[3].
2. FILLED VS. EMPTY TIME. It's a common experience that an interval filled with some activity seems to pass more quickly than an "empty" interval of same duration[4]. Even if several investigations have been made to account for these different ways of experiencing time, the notion of "empty time" is methodologically doubtful.
3. TEMPORAL INFORMATION PROCESSING VS. EXTERNAL TEMPORAL STANDARDS. Some approaches assume that psychological time depends on intra-organism mechanisms such as an *internal clock*[5], a *temporal information processor*[6], *mental models* and so on. Others stress the role of external temporal frameworks or scenarios.
4. PROSPECTIVE VS. RETROSPECTIVE PARADIGM. Block[7] has distinguished two main approaches to the experimental study of time. In a prospective paradigm the subject is previously informed that he will be asked to judge the duration, while in a retrospective paradigm only after the experiment is the subject explicitly requested to judge duration.

In order to capture some aspects of our temporal experience that, as far as I know, have been neglected by above mentioned frameworks, I would like to go deeper into the relationship between our everyday temporal experience and official timing. This approach, that I have named Mental Clock Model, considers our everyday passing duration as the product of an implicit comparison between a hypothetical internal clock and a number of external temporal landmarks (watches, time-tables, "to be on time", "to be late") that make up our *official timing*. The model relies on the assumption that the mental clock tends to modify its speed according to different "contextual" variables such as those identified by Block (e.g. characteristics and activities of the subject; complexity of events; time-related behaviors, and so on).

Our usual questions about time (*what time is it? how long have I being doing this? when will I finish this?* and so on) lead us to realize that we constantly try to synchronize our mental clock to official timing.

Specifically, the mental clock model relies upon the assumption that our mental clock slows down, when we are attending to tasks that need more attentional resources or when we are very busy (see the above mentioned "filled time" notion). On the contrary, if a situation requires less attention or is restful, our mental clock tends to go faster (see the empty time notion). These variations of the mental clock are cognitively impenetrable, i.e. they are not directly perceived by our cognitive system. As a consequence, our conscious perception of time depends upon a continual comparison of our mental time with official timing schedules. This comparison that usually leads us to "watch the time" generates the following counter-intuitive *relativistic effect* (or *inverse time illusion*): when our mental clock goes *slower*, we perceive an *acceleration* of official timing ("time flies"); on the contrary, when our mental clock goes *faster*, we perceive a *deceleration* of official timing ("time hangs heavy"). Moreover, our "watching the time" behavior is quite different in these two temporal conditions. If we are engaged in stressful activities or if events require a lot of attentional resources (filled time), we will tend to disregard the official timing, e.g. watching the time less frequently. As we adjust our subjective time to official timing, our watch will appear to run away and we will realize we are late.

On the contrary, if we are rather inactive in an "empty time" situation, we will probably focus our attention on the passing duration, watching more frequently the time which will appear to go slowly. From this point of view, the claim of attentional models that duration estimation increases with the observer's attention to time could be explained not by temporal versus non-temporal information processing[5], but by an acceleration of our mental clock in comparison with the official timing (in other words, if we are sitting in a waiting-room and we have nothing to do but watch the time, this "empty" interval will produce an increase of speed of our mental clock that will calculate this perceived duration as slower/longer than the official one).

In order to give empirical support to the Mental Clock Model, I am at present investigating some everyday behaviors in relation to our (hypothesized) dual tuning both to the mental time and the official one. Specifically, I'm providing for a "time evaluation questionnaire" including a list of everyday activities and/or situations in which the individuals perceive themselves "to be late" or "to be early". The subjects will be requested to fill it in after having realized to have been "late" or "early". In the former case, we could expect that busy people will *underestimate* time, since their mental clocks are *decelerated* in comparison with the official time; in the latter case, we could expect that rather inactive people will *overestimate* time since their mental clocks are *accelerated* as regards to the official time).

However, the model is incomplete and needs further refinements. For instance, it doesn't explain how the hypothesized comparison between mental and official time could influence not only the perceived duration but also the temporal memory (remembered duration). In spite of its limitations, the Mental Clock Model offers us a cognitive-ecological approach that attempts to incorporate in Block's contextualistic framework our everyday "watching the time" behavior.

REFERENCES

1. Michon J.A., 1990, Implicit and Explicit Representation of Time, *Cognitive Models of Psychological Time*, Block (ed.), Lawrence Erlbaum Associates, Inc., Publishers.
2. Gibson J.J., 1975, Events are perceivable, but time is not, *The study of time II*, Fraser & Laurence (eds.), New York: Springer-Verlag.
3. Zakay D., 1990, The evasive art of Subjective Time Measurement: Some Methodological Dilemmas, *Cognitive Models of Psychological Time*, Block (ed.), Lawrence Erlbaum Associates, Inc., Publishers.
4. Campbell S.S., 1990, Circadian Rhythms and Human Temporal Experience, *Cognitive Models of Psychological Time*, Block (ed.), Lawrence Erlbaum Associates, Inc., Publishers.
5. Treisman M., 1963, Temporal discrimination and the indifference interval: Implications for a model of the "internal clock, *Psychological Monographs*, 77.
6. Shepard R.N., 1984, Ecological constraints on internal representation: Resonant kinematics of perceiving, imaging, thinking, and dreaming, *Psychological Review*, **91** 417-447.
7. Block R.A., 1990, Models of Psychological Time, *Cognitive Models of Psychological Time*, Block (ed.), Lawrence Erlbaum Associates, Inc., Publishers.
8. Cardaci M., Gangemi A., 1993, La stima del tempo: una verifica dell'"Internal Clock Hypothesis", *Comunicazioni Scientifiche di Psicologia Generale*, Edizioni Scientifiche Italiane, **10** 127-140.
9. Migliore M., Messineo L., and Cardaci M., 1999, Modeling the effects of cognitive load on the calculation of elapsed time. Soc. Neurosci. Abs. 25, 1390.
10. Migliore M., Messineo L., and Cardaci M., 1999, Spike latency as coding of elapsed time: a simulation study. The 3rd International Workshop on Neural Coding '99., Osaka, Japan.
11. Thomas E.A.C., Brown I., 1975, Time perception and the filled duration illusion, *Perception and Psychophysics*, **16**.

The perception of time

WOLFGANG HANKE
Universität Hohenheim, Institut für Physiologie, Stuttgart, Germany

1. INTRODUCTION

Any biological organism and especially all human beings depend in their existence on the ability to sense and the environment. To sense the environment means to be able to react to stimuli from the outside of the organism, either physical or chemical stimuli, and to transform these stimuli into electrical signals which can be further processed and interpreted by the brain. For this purpose the organism is equipped with a variety of sensory organs or systems which are specialized in certain modalities[1]. A sensory system can be based on a specific cell-membrane based receptor mechanism or can be unspecified. Although historically also used for complete cells the terminus receptor today more often is used for a cell-membrane incorporated protein-molecule which reacts to the given stimulus by a conformational change and by this triggers further biochemical and electrical events within a cell. These finally induce action potentials, or change frequencies of action potentials, which are transmitted to the central nervous system and processed there. Finally after additional treatment, the stimulus is mentally recognized. The construction of a typical receptor based system is outlined in figure 1. Here the components are classified due to their functional aspects together with the biochemical basis of such functions. Unspecified sensory systems have no receptor, but often the complete cell-membrane fulfils the same task. An example would be the reaction of cell membranes to temperature changes with changes in fluidity, which then results in changes in the electrical properties of the complete membrane and affects action potential sequences. Additionally the question of CNS tissue itself being able to sense certain stimuli directly by having the properties of an excitable medium are more and more discussed.

Studies on the Structure of Time: From Physics to Psycho(path)ology
Edited by Buccheri *et al.*, Kluwer Academic / Plenum Publishers, New York, 2000.

In addition to sense physical and chemical stimuli the organism must also be able in a certain way, to sense space and time in order to arrange reactions to stimuli in the environment. The question whether time and space are separately perceived (sensed) by a human being or whether they are a multidimensional framework in which stimuli and reactions are just located and which itself is re-constructed from the stimuli, is at least partially the content of this text.

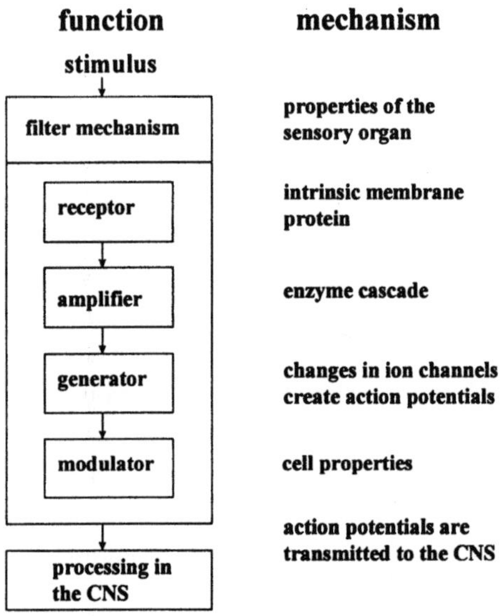

Figure1. Principal construction of a sensory system in a cell, based on a membrane-integral receptor protein. The receptor protein is the entity reacting to a specific stimulus from outside the cell with a conformational change. This conformational change then triggers all following events and actions. Of course changes in system parameters (i.e. temperature) will affect the function of the receptor protein, too, and due to physical reasons at a very low probability even receptor reactions without stimulus are possible. Nevertheless, the receptor is highly specific and widely independent on other inputs and stimuli than the one it is specified for.

2. CLASSICAL SENSORY PHYSIOLOGY

To try to understand whether the human organism senses time, and when it does so, how this might be achieved, a basic classification of the typical human senses it quite useful. As already said, chemical (i.e. an odor) or physical (i.e. light) stimuli can be considered.

- First, any such stimulus transfers a small amount of energy to the sensory system which is necessary to trigger the receptor or induce a change in the cell-membrane. The amount of energy transferred can be very small, possibly even smaller than the limit of thermal noise in case the sensory system is considered to have the properties of excitable media, nevertheless, this amount of energy is finite.
- Second, any such stimulus-sensor system is specific. A sensory system reacts to the stimulus with a high selectivity. Of course wrong stimuli also can trigger a system, for example, a beat on an eye might induce the sensation of stars, but basically the receptor molecule in visual sensory cells, the rhodopsin, is specifically tuned to absorb light quanta. Also, more unspecified systems exist but even these have a certain selectivity, and in any case at least a somehow defined sensory system is given.
- Third, the sensation of each stimulus is independent. The visual cell and thus the eye can react to light even in case all other senses are switched off. Basically the same behavior is true for all other described classical sensations.

Finally, all stimuli after reception have to be transformed to electrical signals, transduced to the brain and further processed there. Even in case the central nervous system (CNS) is narcotized the sensory cells themselves can be still functional and produce electrical signals as a consequence of incoming stimuli. However, under such circumstances no real perception of the stimulus will be given. Related to this last point is the question, whether sensations can be produced in the brain directly, for example, by gravity influences or by magnetic or electrical fields. In such a case the brain itself would function as a sensory organ. This type of behaviour becomes even more reasonable, when assuming that the brain has all the properties of an excitable medium, and thus is extremely sensitive to any small change in its parameters[2,3]. It has been also argued that sensations can be possible under the condition that the temporal correlation between stimulus and sensory event has been lost. Going even further, sensations might be produced in the brain even without any input. Whether this is real perception or just illusion is a more philosophical question which cannot be discussed here in detail.

3. DOES A TIME SENSOR EXIST?

According to the definitions and statements above the question of whether a human subject can perceive time or not, may now be addressed a little bit more in detail. As the nomenclature is not absolutely clear according to the terminus perception, sometimes instead either the terminus sensation or experience will be used.

Unfortunately, even on a very basic physical level it is not really clear yet what physical time is, and of course it is difficult to classify a stimulus when not knowing the nature of this stimulus. The problem becomes even more critical in the view of this text that the physiological basis of temporal experience is also open. Nevertheless, it seems reasonable to classify time as a physical property for our purposes. This physical time, which is different from the so called mental time (see below) and even might not really exist, for our purposes is defined to be free of any meaning, metric, chronological, formal, abstract, quantitative, and independent on any ongoing action. Related to the first argument from above, the physical time by sure has no intrinsic energy which could be transferred to a sensory system or receptor. A first simple statement then would be "*there is no perception of time*"[4].

When we now look for a specified sensory system, a cell or a receptor protein or even a complete organ, which senses time as is, there has also nothing been found up to now in physiological experiments. According to an unspecified perception of time also nothing is known. Due to this lack of information no final statement according to this question can be made.

Let us now ask whether independence is given in the sensation of time. This question is difficult to answer, but at least some information is available. Experimentally, for example, all other senses of a human being would have to be switched of, and later it would have to be asked, whether under these conditions there was still a sensation of time. Some difficulties of such an experiment would be:

- morale and ethic reasons
- the brain must remain functional, thus not all intrinsic stimuli (i.e. heart beat) can be switched of
- as the brain must stay functional it might then itself function as a sensory system, violating by this the request for independence.

Nevertheless, some experiment designed towards the stated direction have been done. In these so called isolation tank experiments all classical senses of an individual are disabled as far as possible and the sensations and experiences of the subjects are later discussed. Some changes in the experience of time under such miserable conditions have been reported, but their interpretation is difficult. Also, after longer isolation subjects typically react psychotic and need psychological help. There is, however, at least some evidence from these experiments that the sensation of time depends on the input of other stimuli, and that possibly a sensation of time would not be given under conditions of total sensory isolation. The real existence of a state of the human brain with no sensory input which according to the above arguments would then be a timeless physiological state, must presently remain open. The findings in isolation tanks would nevertheless at least partially violate the request of independence as stated above. Again the consequence must be to state that there is no real perception of time as the criterion of independence seems to be violated.

Finally we have again to ask what happens when the CNS is narcotized to approach the question whether the brain itself senses time. Two different cases are given. Either after partial CNS narcosis there is the sensation that some time has passed, or in cases of complete narcotized the subject after recovery from narcosis typically asks, when the narcosis will start. Obviously under these circumstances any sensation of time passing was lost. The brain regions necessary for this different behaviour can be partially located by pharmacological approaches and can give a first hint what the sensation of time is, and where the basic mechanisms are located.

The behaviour of the human brain during sleep as an other state of mental activity, is complex. A change in the sensation of time very often is given, but during sleep always time as itself is still recognized, thus after sleeping subjects are principally aware of the fact that some time has passed.

4. TEMPORAL EXPERIENCES OF HUMANS

Even from the above simple argumentation it is already clear that there cannot be a simple perception of time as is defined for other sensory modalities. At least human beings, however, experience time as something crucial important to there live, and within this ongoing time all their activities and reactions to stimuli are arranged. A modified and somewhat extended textbook classification[5,6,7] of the phenomena related to the sensation of time is the following:

- simultaneous events
- not simultaneous events
- sequence of events
- duration of events or between events
- history - presence - future
- course, continuity.

Each of the points of this list again tells us that the experience of time is not really independent but related to the existence of events which of course have to be sensed by other sensory systems. Unaffected by the physiological basis and the underlying mechanisms, the above classification is also related to some typical time intervals. To become aware of whether events are simultaneous or not, about 3 to 5 ms are necessary in the auditory system which has the best temporal resolution of human senses. However, to recognize that two events are not simultaneous does not include to know which was first and which was second. To make this differentiation, more time is necessary, typically 30 to 50 ms. The duration of an event also must exceed a certain value before the event can be classified to have a duration. In the sensory system even a single quantum absorption can be recognized. The physical duration of this event is extremely short, far below a μs. The brain, however, does not recognize the duration of this event but just that it happened. The necessary duration of events to be classified according to their length again ranges from some ms to 10 ms or more.

The question of how a temporal arrangement along the line history-presence-future (Fig 2) is done in the brain is even more complicated and not only related to certain time intervals. Of course we know when something has passed and thus is history. Also we know that presence is now. A clean physical definition would be then that presence has no duration and is just a point in time, however, following the above outlined rules, an event with no duration cannot be absolutely recognized. Consequently the physiological presence must have a certain duration. To approach the physical length of the physiological presence is somewhat easier when investigating for example how most humans experience music and rhythm. Listening to a certain piece of music humans usually recognize the rhythm without any problem over long periods. The single events (tone, sound, oscillation) creating the rhythm are, however, by sure not simultaneous, thus the brain must be able to integrate events over a certain period, treat this period as presence, and extract information from the events which physically were arranged in time but now are arranged in another dimension and stored (somehow) in the CNS. The typical duration of this extended presence is about 3-5 seconds (a length of this interval of up to 1 min. is sometimes discussed) as has been found in a variety of experiments. The capability of the human brain to integrate events in general in time most probably is also limited to about this interval. Without going into details it should be mentioned here that the integration period for presence and the memory, even the short scale memory, are of course different entities. An event having been perceived more than 3 seconds ago then basically has to be classified to be history. Everything within the last 3 seconds can be presence.

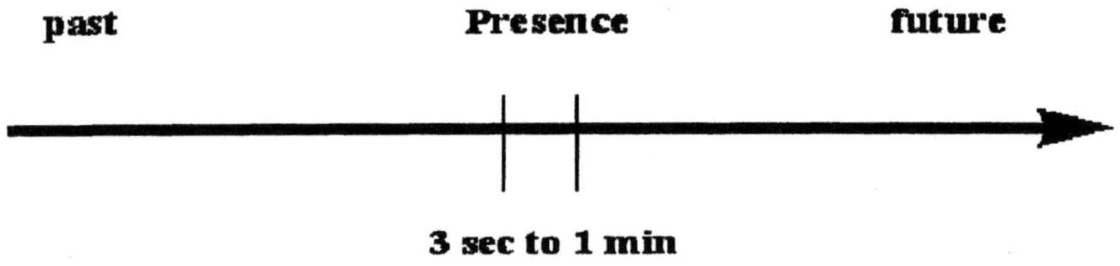

Figure2. Formal representation of the arrow of time and the integration period of brain setting the physiologically defined presence.

However, were the presence is terminated towards the other direction and where the future begins is not immediately obvious from the above statements. Again, a physical definition will be different from a physiological definition. Asides some more esoteric approaches it is a commonly accepted statement in biological and physical sciences that we have knowledge of the history and recognize the presence, but that we are not able to remember the future. Our experience on the other hand is sometimes, that we know what might happen next. This is especially obvious again on a short time scale in listening to music, were we typically are able to predict how the melody and the rhythm should continue, and were we are usually somehow upset when this does not happen as expected. This type of behaviour or sensation, however, is of course no real prediction and perception, but more a prognosis, which might be based on accumulated experiences. The question of whether our "perception" of time ends at the end of what can be defined as presence, must remain open here. Physically, according to the present knowledge, there is no real need for the limitation of only recognizing the presence and remembering the past (neither in the Newtonian mechanics nor in quantum mechanics), although the second law of thermodynamics is sometimes interpreted in giving such a limitation. Also the theoretical approach of the complete universe to be at least up to a certain degree comparable to an excitable medium with self-organizing and evolving structures does not really forbid the recognition of future. There is however, an interesting philosophical background given for this question. In case there would be a real (long-term) perception and thus predictability of future, this universe would automatically be completely deterministic. Of course such a deterministic universe including a perceivable future cannot be ruled out presently, however, for a lot of humans it seems to be no really acceptable (philosophic and religious) perspective.

Related to the question of how history, presence and future are perceived is, to explore how course and continuity physiologically are constructed. In case a physical arrow of time does exist, things might be a little bit more simple. Perception would be able to just follow the physical reality along this arrow. The physical course of time physiologically would be cut into intervals during which physiological integration in the CNS is possible and given, and these intervals would be cued up and mapped to cognition by other mechanisms. Here the physiological time would then be constructed which is of course not metric, not independent on ongoing events, not free of meaning, and not abstract as a possible physical time should be (see above). The physiological time is depending on the physiological and mental status of the individual and it is not homogenous in its course. Interestingly is additional that the mental time obviously is depending on the age of a subject.

Unfortunately, again, the question whether an arrow of time exists physically is still open. Only recently in elementary particle-physics a violation of the basic symmetries has been found, and it is speculated presently, whether this break of time symmetry is sufficient to introduce an arrow of time for the complete universe. Independent on the real existence of a physical arrow of time, the physiological (mental) experience of time passing by must not necessarily be related to any physical mechanism, and the physical and physiological arrow of time most probably are of completely different nature.

Asides the problems with a physical definition of time and the fact that physical and mental time may be different entities there is a long going philosophical tradition to approach to the question whether time can be perceived or exists per se[8,9]. This philosophical treatment of time is at least partially independent on the question whether a sensory system for time exists. It is, however, widely discussed, for example already by Kant, whether time in itself exists and how time is related to other physical and mental entities.

5. PHYSIOLOGICAL CLOCKS

Up to now two things have been stated in this text; first, a real perception of time most probably does not exist, and second, independent on this there is the experience of time by human beings according to the different modalities discussed above. A mechanism must thus exist in the human brain to construct a temporal experience, most probably related to other sensory inputs. A comparison of physiology to technical reality immediately tells us how a time measuring system might be constructed. All physical and technical clocks are measuring time by comparing the temporal organization of cued events to a given oscillator with a know frequency and phase. The most simple assumption for a physiological construct is to request such an oscillator in the CNS and to assume that the temporal organization of events is measured relative to this physiological clock (Fig 3). All other temporal reality in the brain can than be reconstructed from such a measurement.

The variability in mental time in such a construct would be due to changes in the frequency of the reference oscillator and due to further processing in the CNS, after the basic time measurement has been done. The possible organization and localization of such physiological clocks will be discussed later in some more detail. The presence of such a clock would result, however, in some predictions about experiments which can be verified or falsified. The length of the period of most clocks among other parameters is depending on the temperature of the system[10] where the clock is located, and indeed it was for example found, that subjects suffering from high fever have a somewhat accelerated time experience compared to the experience at normal body temperature.

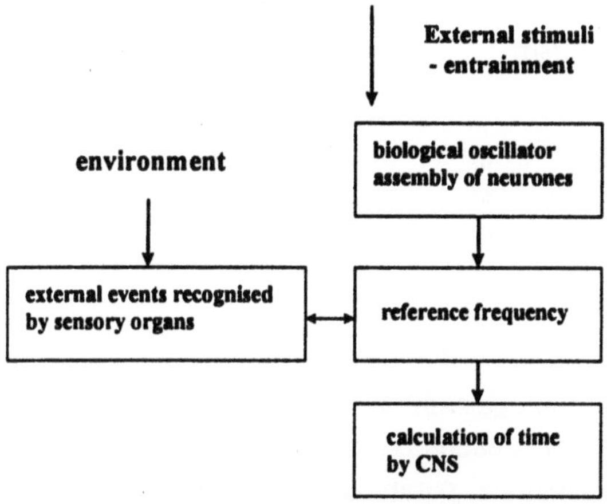

Figure 3. Basic construction of a biological time measurement. The biological oscillator delivers the reference for the time measurement itself. The frequency (or rhythm) of this oscillator of course is dependent on internal and external parameters of the human organism.

Also the frequency of the basic physiological clock must be sufficiently high to measure the shortest time intervals which can be resolved by the human brain, the period should thus be in the ms range. In reality, action potentials in neurons can be shorter than 1 ms and have a frequency up to about 1 kHz, thus a sufficient resolution on this basis would be possible when taking the above given numbers about the recognition of events not being simultaneously into account. By proper coding, this resolution could be enhanced even a little bit further. When the physiological clock is mechanistically based on the behaviour of neurons and the existence of action potentials[11], it should be also depending on the addition of certain drugs changing the properties of neurons and the duration and frequency of action potentials. Experiments addressing this question have been done and indeed it was found that the temporal sensation of subjects can be influenced in a wide range by the application of drugs. Anyhow, it is known that the experience of time is correlated to the actual physiological and mental status of the body (the brain), time can fly or pass extremely slowly, and any such actual state of the mental-body-system is mainly defined biochemical parameters. All these findings indicate that a biological clock exists. The narcosis experiences mentioned above then indicate that this clock indeed is located in certain areas of the CNS.

According to the wide range of human temporal experiences ranging over at least 10 orders of magnitude, a list is given in table 1, and due to the construction of the human organism, it seems in addition to be necessary to have not only one physiological clock but either one clock producing different frequencies, or different clocks with different frequencies. Of course these clocks must be synchronized somehow for proper function of the complete organism.

Table 1. Typical time intervals of human temporal experiences.

1-3	ms	smallest resolvable duration
10-30	ms	time necessary to verify temporal order
3-5	sec	integration period for presence
24	hours	day-night rhythm
28	days	female period
1	year	summer-winter rhythm

In a next step the question must be addressed, whether the physiological clock is running independently and with a frequency defined only by the physical and biochemical parameters of the CNS. The oscillator being the mechanistic basis for this clock in such a case would be free running. Alternatively it could be forced and locked to an external input. The internal input of an externally forced oscillator could be given by any sensory system of the human organism. According to the circadian organization of a lot of human temporal experiences, so called free running experiments are of some interest. In such experiments subjects are isolated from the natural day-night rhythm and create their own light-dark, activity-rest environment. Typically the rhythm set in such experiments by the subjects is about 25 hours, indicating that the physical clock in this interval range has about a period of 25 hours and under natural conditions is forced (entrained) to a somewhat shorter period of 24 hours by the natural day-night sequence mainly via the visual sensory system. It is also worth to mention that after to long exposure to free-run subjects frequently desynchronize and loose any rhythm. This indicates that physiological clocks are not really independent on other sensory inputs[12,13].

The present knowledge about biological systems delivers strong evidence that one of the basic physiological clocks in the human CNS is located in the suprachiasmatic nucleus. This oscillator is most probably responsible for the daily rhythm and 25 hours awake-sleep circadian behaviour of humans as discussed above already. Other biological rhythms are either related to this endogen oscillator or are located at other oscillators in the CNS.

6. CONCLUSIONS

There is no specified or unspecified receptor system for the perception of time known in human beings, not even in the CNS under the extended view of the brain being a sensory system by itself. Accordingly a real perception of time comparable to the perception of other sensory modalities seems not to exist in humans.

The experience of time is most probably based on mechanisms comparable to event counters or frequency measuring devices and seems to be located at different places in the CNS. These endogen oscillators are most probably constructed from assemblies of neurons. The real basic molecular (and cellular) mechanisms of the physiological clocks are still not known, but even simple chemical oscillators as for example the BZ-reaction[14] can create rhythmical behaviour.

There are some similarities in the perception of time and the perception of space, which may be related to the physical relation of time and space. However, the statement that the sensation of time is essentially coupled to changes or displacements in space cannot be verified form the presently available information.

The time physiologically experienced and the physical time, in case it exists, are completely different modalities. A possible perception of future events, as is still thinkable on physical reasons, in a strict interpretation would lead to a completely deterministic universe. This is still possible on the basis of given physical (cosmological) theories, but would have the strong philosophical implications.

REFERENCES

1. Fain G.L., 1999, *Molecular and cellular physiology of neurons*, Harvard University Press, Cambridge, MA, USA.
2. Kelso J.A.S., 1995, *Dynamic patterns: The self-organisation of brain and behaviour*, MIT Press, Cambridge, MA USA.
3. Prigonine I. and Stengers I., 1985, *Order out of chaos*, Flamingo Press, London, UK.
4. Jordain R., 1998, *Das wohltemperierte Gehirn*, Spektrum Akademischer Verlag, Heidelberg, Germany.
5. Sinz R., 1980, *Chronophsychophysiologie, Chronobiologie und Chronomedizin*, Akademie Verlag, Berlin, Germany.
6. Pöppel E., 1988, *Mindworks: Time and conscious experience*, Harcourt Brace Jovanovich, Boston, MA, USA.
7. Pöppel E. and Schill K., 1995, Time perception: problems of representation and processing, *Handbook of brain theory and neural networks*, MIT Press, Cambridge, MA, USA, 987-990.
8. Heidegger, 1928, *Sein und Zeit*, Max Niemeyer, Hamburg, Germany.

9. Husserl E., *The phenomenology of internal time consciousness*, (translated by Churchill, J.S.), Indiana University Press, Blomington, USA.
10. Pavlidis T., 1973, *Biological oscilaltors: Their mathematical analysis*, Academic Press, New York, USA.
11. Goldbeter A, 1996, *Biochemical oscillations and cellular rhythms*, The molecular basis of periodic and chaotic behaviour, Cambridge University Press, Cambridge, UK.
12. Winfree A.T., 1980, *The geometry of biological time*, Springer Verlag, Berlin, Germany.
13. Winfree A.T., 1987, *Zeitstruktiuren des Lebendigen*, Spektrum Akademischer Verlag, Heidelberg, Germany.
14. Zaitkin A.N. and Zhabotinski A.M., 1970, *Nature*, Concentration wave propagation in two-dimensional liquids-phase self-oscillating systems, **225** 535-537.

Subjective Time Versus Proper (Clock) Time

[1]RONALD P. GRUBER, [2]LAWRENCE F. WAGNER, AND
[3]RICHARD A. BLOCK
[1]*Stanford University, Stanford, CA, U.S.A.*
[2]*DSP Technologies, Oakland, CA, U.S.A.*
[3]*Montana State University, Bozeman, MT, U.S.A.*

1. INTRODUCTION

A few physicists have recently made some radical proposals about the nature of physical time. These theories, in turn, relate to proposals that psychologists have made and evidence that they have collected about the nature of subjective time. Exploring these interdisciplinary connections may be fruitful.

2. HYPOTHESES FROM PHYSICS

A few physicists have recently proposed that the appropriate measure of time is not necessarily clock time (or *proper time*, as they call it), but instead the length of time it takes to process one bit of information. Working separately, Dyson[1] and Tipler and Barrow[2] each made two assumptions about time when predicting the consequences to humans as the universe evolves: (1) Subjective time is based on the overall information-processing rate, and (2) the subjective experience of life's duration is related to (if not based on) the total information processed. Consequently, life's duration and longevity can be measured in terms of either a person's chronological age or the total information the person has processed.

Physicists have discussed two main theories about the evolution of the universe in space (and time), the open and closed cosmologies. In the open-cosmology theory, the universe will infinitely continue to expand. As a

Studies on the Structure of Time: From Physics to Psycho(path)ology
Edited by Buccheri *et al.*, Kluwer Academic / Plenum Publishers, New York, 2000.

result, the temperature of the universe will approach absolute zero. In the closed-cosmology theory, the present expansion of the universe will reverse, and the universe will collapse. As a result, the temperature of the universe will approach infinity. (We will not discuss a third, arguably less likely scenario, in which the present expansion of the universe will gradually decrease to zero, resulting in neither an infinite expansion nor a gradual collapse.) Although current evidence seems to support an open-cosmology theory more than it does a closed-cosmology theory, future evidence may alter that view.

2.1 Open Cosmology Theory

Based on the open-cosmology theory, in which the universe will forever continue to expand, Dyson[1] predicted that as the temperature of the universe approaches absolute zero, any sentients will experience a dramatic decrease in their information-processing rate. Consequently, time will seem to pass so rapidly for them that a year might seem like minutes. Sentients living under these conditions will experience their life to be very short compared to their chronological ages. Because there is a finite amount of energy for a sentient civilization, Dyson suggested that sentients should hibernate in order to prolong their chronological ages. This way at least a sentient's chronological age, if not the experienced lifespan, could be increased.

Dyson calculated the change in subjective time as the temperature of the universe decreases using the following function, assuming a scaling law in an open cosmological model:

$$u(t) = f\int_0^t \theta(t')dt'$$

where $u(t)$ is subjective time, f is a scale factor, and $\theta(t')$ is the temperature of the sentient. We note, however, that this equation applies only to a conglomerate of sentients, not to any individual sentient. As the temperature decreases to near absolute zero in an open universe, sentients of the future will have to adapt. It is assumed that efforts to maintain their normal body temperature will ultimately be futile. They may have to replace all their bodily parts with temperature-resistant materials if they do not wish to freeze. From a physics viewpoint, if not a biological viewpoint, this scenario is possible[3].

2.2 Closed Cosmology Theory

In contrast, a closed-cosmology theory proposes that the universe will not expand forever, but will collapse. Based on this assumption, Tipler and Barrow[2] predicted the opposite of what Dyson[1] predicted: As the universe collapses and its temperature approaches infinity, sentients will experience a dramatic increase in their information-processing rate and consequently in their subjective time. As the information-processing rate approaches infinity in the last three minutes of the universe, sentients will be processing information at such a high rate that their subjective time will be infinitely long. The experienced lifespan for these sentients will be enormous, completely out of proportion to their chronological ages. Sentients living in the last three minutes will therefore experience immortality (infinite longevity) as the universe ends.

Tipler and Barrow also suggested that as the information-processing rate of future sentient beings decreases relative to proper time, the contrast between proper time and subjective time will be striking. As a result of the discrepancy between proper time and subjective time, sentients presumably will not hesitate to reject proper time as the appropriate measure of time. In the current astrophysical environment, proper time is presumably accepted as the appropriate measure of time because the information-processing rate of humans is directly proportional to proper time. In other words, one second of subjective time corresponds fairly well with one second of proper time.

The implication of these two physics hypotheses was profound in that physicists were suggesting that proper time is not the only important time in physics and biology. For what was arguably the first time, physicists were beginning to acknowledge that subjective time is important, although perhaps not until sentients' information-processing rate is drastically altered. Although the open- and closed-cosmology scenarios are speculative and their predictions are opposite, the assumptions about subjective time are similar.

3. EVIDENCE FROM PSYCHOLOGY

Guyau[4] and James[5] were two of the first theorists to relate time experiences to the concept of human information processing in an attempt to explain phenomena of subjective time. Although neither of them used the term *information processing*, they discussed the influences of the number of stimuli, the intensity of stimuli, differences between stimuli, and other such information-processing considerations. They both thought that subjective time is not determined so much by clock time as by the events people experience.

In mathematics and physics, the number of bits processed per second is a system's information-processing rate. In psychology, however, most evidence indicates that chunks (meaningful units) instead of bits (mathematical units) are critical in human information processing and memory[6]. However, for our purposes, we can conveniently ignore the difference between bits and chunks. One of the crudest but simplest ways to measure human information processing is by observing the time it takes a person to react to a stimulus or to a series of stimuli. The shorter the latency of the person's response or the more bits (or chunks) a person processes during a time period, the faster is the person's information-processing rate.

Because the physics hypotheses deal with the effects of catastrophic changes in information-processing rates, they may seem to have no relevance to human beings in this epoch. However, it has been known for many decades that such factors as temperature and age influence the human information-processing rate. For this reason, there is a parallel to what the physics hypotheses propose about changes in sentients' subjective time and subjective lifespan as the universe evolves. Evidence from studies of subjective time may clarify the physics hypotheses.

3.1 Temperature, Information-Processing Rate, and Subjective Time

Some early evidence seemed to link a person's subjective time and his or her information-processing rate. Hoagland[7,8] reported that he went to the drugstore to get some medicine for his wife, who had a fever of 104° F. He was gone for 20 minutes, yet on his arrival his wife insisted that he was gone for 30 minutes. Since then, more controlled experimentation has revealed that a person's experience of duration tends to lengthen as body temperature increases (see, for example, Hancock[9]). In addition, deep-sea divers whose body temperatures drop precipitously return to the surface claiming that they had been underwater for much less time than the clock actually stated[10]. Siffre[11], a young geologist, lived for two months alone in a cave on a floor of melting ice in near-freezing conditions. He suffered from near-hypothermia, a condition of semi-hibernation, and experienced time to pass very slowly. He estimated that he had been in the cave for only about four weeks. Decreased body temperature also leads to a decrease in a person's information-processing rate[12].

Subjective time estimates can be made prospectively or retrospectively, however, and these two kinds of estimates may differ. When tested prospectively (i.e., when subjects know in advance that they will make time estimates), people with decreased body temperatures do not show decreased time estimates for target durations of one hour[13]. When tested retrospectively, however, duration judgments decrease when body temperature falls by as little as 1.0 F. [14].

3.2 Age, Information-Processing Rate, and Subjective Time

Since ancient times, it has been commonly accepted that subjective time changes with age. Rosenberg[15], a neurologist, theorized that the seemingly more rapid passing of time as a person ages may be attributed, at least in part, to changes in the person's information-processing rate. The effect that age has on the information-processing rate has never been in question. For a variety of information-processing tasks (verbal and visuospatial), the mean information-processing rates increase from birth and reach a maximum[20] at about age 20. They decline from that point on, as Lawrence, Meyerson, and Hale[17] found (see Figure 1). The processing score at any age is best appreciated by expressing it as a percentage of that noted at about age 20, the age of maximum performance. At age 10, the mean information-processing rate is about 61% of what it will be at age 20. At age 60, the verbal information-processing rate drops to 78% of that value and the visuospatial information-processing rate drops to 38% of that value.

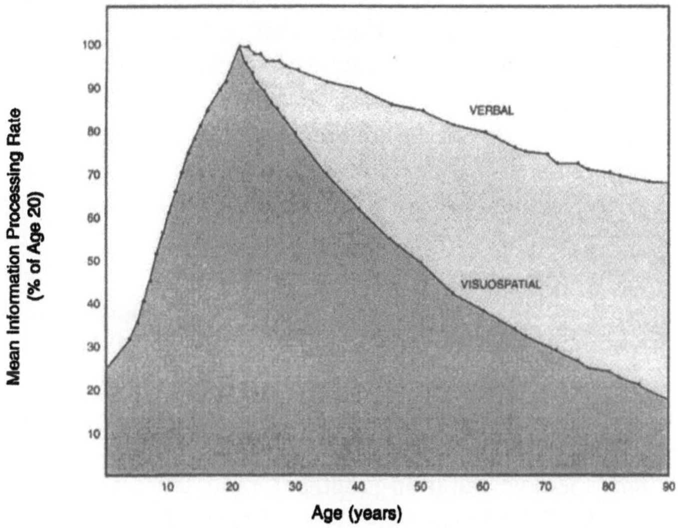

Figure 1. For a variety of information-processing tasks (verbal and visuospatial), the mean information-processing rate increases from birth and reaches a maximum at about age 20. The processing rate at any age is best appreciated by expressing it as a percentage of that at about age 20, the age of maximum performance. Data are taken from Kail[34] and Lawrence et al[17].

The literature on age-related changes in subjective time, however, has been more controversial. Although many psychologists have asserted that physical time seems to accelerate as a person ages, specific studies to measure the extent of that acceleration (in terms of years) did not begin until relatively recently. Lemlich[18] asked people of various ages to indicate how much slower or faster the year seems to go by at the present time than when they were approximately one-half their present age. Almost all adults indicated that time seems to pass faster now than it did during a former period in their lives. Furthermore, middle-age adults report that time seems to pass more quickly than young adults do.

Because the estimation of years is difficult and subject to many idiosyncrasies, most investigators of subjective time have looked at shorter durations—that is, on the order of seconds and minutes. In two meta-analyses, Block, Zakay, and Hancock[19,20] noted that prospective time estimates change from age 10–72. In Figure 2, we have combined their findings, with the duration-judgment ratio (the ratio of the subjective to objective time) plotted as a function of chronological age. The results varied depending on the method of testing. For example, using the verbal estimation method (in which a person estimates a duration in numerical units such as minutes and seconds), the duration-judgment ratio decreases from age 10 to about age 20–25, but increases again to about age 70. These data reveal a nonmonotonic (U-shaped) function over the lifespan, which appears to be the opposite of what Rosenberg[15] had predicted. However, the findings probably reflect the opposite effects that are often obtained in the prospective and retrospective paradigms[21]; these data all came from a prospective paradigm.

Retrospective time-estimation studies show different results. Only three researchers have reported the results of studies on retrospective duration judgment as a function of age (see Figure 3). Using a reproduction time-estimation method (in which a person delimits a subsequent duration to be subjectively equal to a previously experienced one), Kelley[22] and Vanneste and Pouthas[23] both found that the duration judgment ratio decreases from about 0.78 at about age 20 to 0.47 and 0.58 at age 65–83. Using a different method, Herman, Norton, and Roth[24] studied time estimation in young children and adolescents. Although they did not find an age effect in one of their studies, they did find a significant effect in the other: The retrospective duration-judgment ratio increased from 0.44 for 9-year olds to 0.51 for 19-year-olds. Thus, these retrospective results agree with Rosenberg's[15] hypothesis.

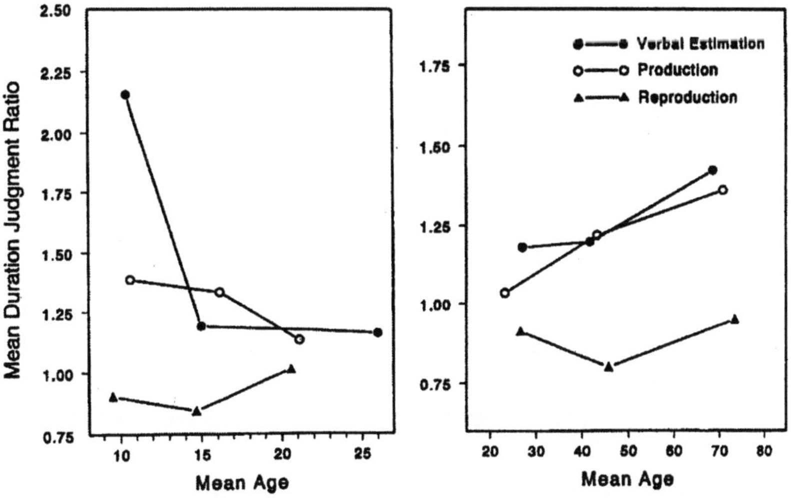

Figure 2. In two meta-analyses, Block and Zakay (1998, 1999) noted the change in prospective time estimation from age 10-72. The duration-judgment ratio (the ratio of the subjective to objective time) is plotted as a function of age. The results varied depending on the method of testing.

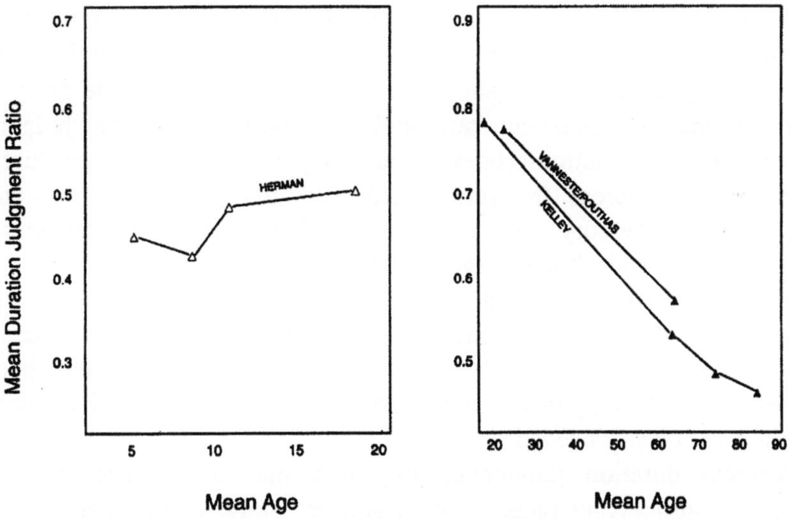

Figure 3. Using a reproduction time-estimation method, Kelley (1980) and Vanneste and Pouthas (1995) found that the duration-judgment ratio decreases during the adult years. Using a comparison method, Herman et al. (1983) found a significant age effect during childhood and adolescence.

Why do prospective and retrospective studies show opposite results? First, some prospective studies do show results similar to retrospective studies. Using target durations during which subjects were busy, Craik and

Hay[25] (see also Mangan, Bolinskey, and Rutherford[26]) found that the duration-judgment ratio decreases with aging. Therefore, the choice of paradigm is an important factor. Second, prospective and retrospective duration judgments involve inherently different mechanisms, with prospective judgments mainly influenced by attention and retrospective judgments mainly influenced by memory. For example, decreasing the processing load (e.g., the number of stimuli presented during the target duration) decreases retrospective duration judgments but increases prospective duration judgments[21]. It is not surprising to see opposite results when processing rates decline.

3.3 Information Processing versus Duration Judgment

The retrospective data suggest that information-processing rates may be involved in the developmental and aging changes in retrospective duration judgment. In order to compare duration-judgment data and information-processing data, it is necessary to compare the duration-judgment ratio at any given age as a percentage of that at age 20, the age of presumed maximum performance. Although data from only three studies is meager, it is apparent that the age-related change in retrospective duration judgments (shown in Figure 3) resembles that noted for overall information-processing rate (shown in Figure 1). On either side of age 20, both information-processing rate and retrospective duration judgments decrease. The information-processing rate may mainly reflect perceptual and motor speed, whereas the retrospective duration judgments may more heavily reflect memory processes. Nevertheless, there may also be some similarity in the underlying causes of the age-related changes.

No single hypothesis can account for all developmental and aging differences in duration judgments. Nevertheless, the information-processing rate may be an important factor to consider, particularly in the retrospective studies. More developmental and aging studies of retrospective duration judgment are clearly needed.

Whereas duration judgments may turn out to correlate well with information-processing rates in these studies, body temperature does not. Changes in information-processing rates with aging are probably more a result of brain deterioration[27] than a brain temperature decrease. Furthermore, body temperature (and basal metabolism) does not peak at age 20; it decreases monotonically across the lifespan[28]. Thus, changes in subjective time over the lifespan are not likely to be solely a result of body (i.e., brain) temperature changes.

4. PHYSICS VERSUS PSYCHOLOGY: COSMOLOGICAL VERSUS CURRENT ENVIRONMENT

Now consider subjective time in the cosmological situation versus subjective time in the current astrophysical environment. We note that the physicists' hypotheses concern subjective time under extreme cosmological environments—that is, near the end of the universe. According to the physics hypotheses, subjective time in the cosmological situation is related to the total amount of information processed, and the perception of that information is presumed to be proportional to the perception of time. In contrast, according to both Rosenberg[15] and most psychologists, subjective time entails the perception of time itself, however well it correlates with the information that is processed.

In both the open-cosmology scenario and the current astrophysical environment, a shortening of subjective time may be attributed (at least in part) to a decreased information-processing rate. However, the etiology and rate of the information-processing decrease is quite different between the two. In the scenario of the open cosmological environment, the processing rate declines drastically due to a drastic temperature decrease. In the current environment, the processing rate declines with aging (after approximately age 20) due to a slight decline in brain function. In both situations, subjective time (at least the retrospective aspect) may be related to the overall information-processing rate.

4.1 Recalibrating versus Abandoning Subjective Time

It is important to examine a few issues that relate to the assumptions and predictions of the physics hypotheses. One prediction is that humans will not hesitate to reject proper time as the appropriate measure of time when it decelerates relative to subjective time[2]. However, it is far from certain that this will be the case. Currently, humans do not abandon proper time for subjective time despite the fact that short-term retrospective time estimation declines slightly with aging. Perhaps that is because of the variability and malleability of subjective time. However, another process must be considered: the continual recalibration of subjective time to proper time. In estimating durations, one compares some currently accumulated temporal "pulse count" with pulse count totals (stored in a reference-memory store) that accumulated during earlier, similar time periods[29,30]. Therefore, if there is a gradual change in the rate of a pacemaker, which is producing the pulses (perhaps because of a gradual brain temperature change), people experience and judge a duration much the same way they did before the gradual change.

Because many factors (e.g., boredom) influence subjective time, people often erroneously regard subjective time as an illusion, whereas they regard the time of physics (or of clocks and watches) as "real". However, humans might not abandon subjective time if they recognized that one of the basic mechanisms on which it is founded, the information-processing rate, is "real". In any event, it is questionable that sentients living in an extreme astrophysical environment (such as that existing in the very cold stages of an open universe) will reject proper time in favor of subjective time. Sentients may instead recalibrate their subjective time to adapt to the gradual changes in their timekeeping processes and rates.

On the other hand, a sudden change in the rate of a pacemaker ("pulse rate") will lead to a markedly altered time experience. Binkofski and Block[31] (see also Block and Zakay, this volume) described the case of a man with a brain tumor which apparently slowed his internal clock and, hence, his information-processing rate. The effects were surprisingly sudden and dramatic. He experienced normal environmental events as occurring very rapidly, much like a videotape played on fast forward. He could not even tolerate watching television, because the progression of events was too quick for him to follow. Perhaps because his information-processing rate was suddenly and drastically slowed, he also greatly misjudged a one-minute duration. A similar change in sentients' information-processing rates will probably not occur in an astrophysical environment characterized by gradual cooling. The notion that sentients living in the very cold stages of an open universe will reject proper time assumes that they will not be able to recalibrate to compensate for the gradual changes.

4.2 Measuring Life's Duration by Information Processed

Dyson's[1] and Tipler and Barrow's[2] second assumption is that the subjective experience of life's duration can be measured either by the chronological age or the total amount of information processed (one's cumulative number of thoughts). They suggested that sentients may prefer to measure life's duration by the subjective duration of their life which, in turn, is reflected by the amount of information processed in their lifetime. For example, in the closed cosmology scenario, a sentient could conceivably be 10 years old and yet have processed more information than someone in this epoch who has lived 100 years. Chronological age for that sentient would be a less meaningful reflection of life's duration than the total bit or chunk count. Proper time would not be a good scale of time for what the sentient subjectively experiences.

If we stop to wonder where this notion came from we do not have to look any further than the automobile. As opposed to people, cars are judged not only by the year they were manufactured but also by their mileage. The mileage of a car would seem to be the counterpart to the total information processed by humans—what could be called "cerebral mileage." Although it is difficult to measure the cerebral mileage in humans, in principle it is possible.

To estimate the amount of information processed in a lifetime, we define the total information processed in the 20th year as one bit year. Then, if we know the information-processing rate for each year (by taking the mean of the visuospatial and verbal information-processing rates, for example), we can calculate the yearly bit count. Although the yearly bit count declines each year after age 20, the cumulative bit count (called the bit age) increases with aging (see Figure 4). Because the yearly bit count declines after age 20, the bit age does not equal the chronological age. In fact, the bit age of the average 60 year-old is about 50 bit-years. In essence, the cerebral mileage of a 60 year-old is less than it would have been had his or her information-processing rate not declined over time. Unfortunately, we are far from being able to make these calculations for individuals.

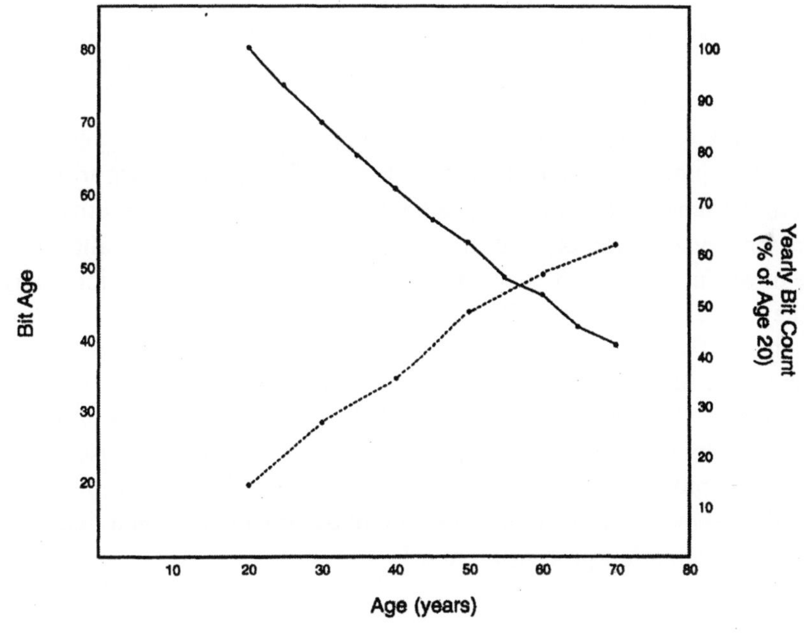

Figure 4. The total information processed in the 20th year is defined as one bit year. Based on the information-processing rate for each year (e.g., the mean of the visuospatial and verbal information-processing rates), one can calculate the yearly bit count. Although the yearly bit count decreases beginning at about age 20, the cumulative bit count (called the bit age) increases with age. The bit age does not equal the chronological age. In fact, the bit age of the average 60 year-old is about 50 bit-years.

Curiously, people do not attempt to estimate their cerebral mileage. In part, this is because they cannot measure their own information-processing rate, and because they certainly cannot calculate their lifetime bit count. They can and do, however, recognize that they slow (both physically and mentally) with age. Instead of estimating the length of their life by summing their activities, people choose to compare their current activity level to that when they were younger or to someone else who is younger. Thus, a 60 year-old may feel that his reflexes are as quick as that of a 50 year-old.

4.3 Two Ways to Prolong Longevity

The most profound prediction of the physics hypotheses is that human longevity may be increased by directly manipulating the information-processing rate. Davies[32] summed it up by proposing that there are two ways to prolong the longevity of sentient beings. One way is to survive as long as possible—i.e., to increase proper time or chronological age. The other way is to look for a means to speed up thinking and experiencing—that is, to increase the information-processing rate. Thus, the physics hypotheses imply that physicians should not only be looking for ways to prolong the chronological age of humans but also for ways to increase their information-processing rate.

Currently, there are no dramatic ways to increase the information-processing rate. Drugs such as caffeine are known to increase the information-processing rate[33], as well as lengthen duration estimates. The effect, however, is minor, and side-effects preclude its practicality. Other, more powerful drugs such as metamphetamine produce more major changes, but they are also dangerous. Perhaps the only safe and successful treatment, although difficult with which to comply, is that of dietary restriction. A substantial reduction of one's total caloric intake (up to 40%) has repeatedly been shown to increase chronological longevity in animals and some humans. Along with that improved longevity, an increase in the information-processing rate has also been observed. Animals on dietary restriction show shortened latencies in maze learning, which reflects an increased information-processing rate[34].

5. SUMMARY AND CONCLUSIONS

The notion that subjective time is based on information processing, first introduced over a hundred years ago by philosophers and psychologists, has in recent decades received support from some physicists. A physics hypothesis that was developed to deal with a cosmological scenario was predicated on two assumptions: (1) subjective time is based on the overall information-processing rate, and (2) the subjective experience of life's duration is related to (if not based on) the total information processed. As a result, life's duration and longevity can be measured either by chronological age or the total information one has processed. In the open-cosmology model, the decline in the temperature of the universe is expected to cause a decline in sentient information processing rates such that there is a drastically fast passage of time and abbreviated longevity.

The physics prediction had no apparent relevance for humans in this epoch. However, a psychology hypothesis suggested that an age-related (non-temperature) induced decline in human information-processing rate is at least partly responsible for the much milder "quick passage of time" that is associated with aging. In an attempt to explore the psychology hypothesis, we analyzed the available data involving age-related information-processing rates and time estimation (duration judgment) studies. Our review provides some support for the association between retrospective duration judgment and the overall information-processing rates.

We also discussed one of the most profound implications of the physics hypothesis, that human longevity can be increased not only by prolonging proper time (chronological age) but also by increasing the human information-processing rate. Whether or not it will be possible to increase life's duration (and thereby longevity) without increasing chronological age remains to be seen. It stands as an exciting challenge for future research.

ACKNOWLEDGEMENTS

We thank Sandra Hale and Joel Myerson for supplying data (which we used in Figure 1). We also thank Peter Hancock for commenting on a draft of this chapter.

REFERENCES

1. Dyson F.J., 1979, Time without end: Physics and biology in an open universe, *Reviews of Modern Physics*, **51** 447-460.
2. Tipler F.J. and Barrow J.D., 1986, *The anthropic cosmological principle*, Oxford, England, Clarendon Press.
3. Krauss L.M. and Starkman G.D., 1999, The fate of life in the universe, *Scientific American*, **281** 58-65.
4. Guyau J.M., 1988, La genèse de l'idée de temps [The origin of the idea of time], *Guyau and the idea of time*, Michon, Pouthas, and Jackson (eds.), Amsterdam, North-Holland, (Original work published in 1890), 37-90.
5. James W., 1890, *The principles of psychology*,Vol. 1, New York, Holt.

6. Miller G.A., 1956, Human memory and the storage of information, *I R E Transactions on Information Theory*, IT-2, 129-137.
7. Hoagland H., 1933, The physiological control of judgments of duration: Evidence for a chemical clock, *Journal of General Psychology*, **9** 267-287.
8. Hoagland H., 1966, Some biochemical considerations of time, *The Voices of Time*, Fraser (ed.), New York, Braziller, 312-329.
9. Hancock P.A., 1993, Body temperature influence on time perception, *Journal of General Psychology*, **120** 197-216.
10. Baddeley A.D., 1966, Time estimation at reduced body temperature, *American Journal of Psychology*, **79** 475-479.
11. Siffre M., 1963, *Hors du temps* [Beyond time], Paris, Julliard.
12. Rammsayer T.H., Bahner E., and Netter P., 1995, Effects of cold on human information processing: Application of a reaction time paradigm, *Integrative Physiological and Behavioral Science*, **30** 34-45.
13. Aschoff J., 1998, Human perception of time intervals, *Journal of Biological Rhythms*, **13** 437-442.
14. Campbell S.S., Murphy P.J., and Boothroyd C.E., 1999, *Relationship between long-term time estimation and the circadian course of body core temperature*. Manuscript submitted for publication.
15. Rosenberg R.N., 1979, Inverse relationship of the velocities of perceived time and information processing events in the brain: A potential bioassay for neural functions: A hypothesis, *Medical Hypotheses*, **5** 1305-1312.
16. Kail R., 1991, Developmental change in speed for processing during childhood and adolescence, *Psychological Bulletin*, **109** 490-501
17. Lawrence B., Meyerson J., and Hale S., 1998, Differential decline of verbal and visuospatial processing speed across the adult life span, *Aging Neuropsychology and Cognition*, **5** 129-146.
18. Lemlich R., 1975, Subjective acceleration of time with aging, *Perceptual and Motor Skills*, **41** 235-238.
19. Block R.A., Zakay D., and Hancock P.A., 1998, Human aging and duration judgments: A meta-analytic review, *Psychology and Aging*, **13** 584-596.
20. Block R.A., Zakay D., and Hancock P.A., 1999, Developmental changes in human duration judgments: A meta-analytic review, *Developmental Review*, **19** 183-211.
21. Block R.A. and Zakay D., 1997, Prospective and retrospective duration judgments: A meta-analytic review, *Psychonomic Bulletin and Review*, **4** 184-197.
22. Kelley N.L., 1980, Accelerated time among the aged: A memory hypothesis (Doctoral dissertation, University of Florida, 1980), *Dissertation Abstracts International*, **41** 2365B.
23. Vanneste S. and Pouthas V., 1995, Estimation temporelle prospective et rétrospective chez la personne âgée: Comparaison avec le jeune adulte [Prospective and retrospective time estimation in elderly people: Comparison with young adults], *Bulletin de Psychologie*, **48** 539-543.
24. Herman J.F., Norton L.M., and Roth S.F., 1983, Children and adults' distance estimations in a large-scale environment: Effects of time and clutter, *Journal of Experimental Child Psychology*, **36** 453-470.
25. Craik F.I.M. and Hay J.F., 1999, Aging and judgments of duration: Effects of task complexity and method of estimation, *Perception and Psychophysics*, **61** 549-560.
26. Mangan P.A., Bolinskey P.I., and Rutherford A.L., 1997, Underestimation of time during aging: The result of age-related dopaminergic changes?, *Society for Neuroscience Abstracts*, **23** 2003.

27. Cerella J., 1990, Aging and information-processing rate, *Handbook of the psychology of aging*, Birren and Schaie (eds.), San Diego, Academic Press 201-221.
28. Altman P.L. and Dittmer D.A., 1968, Metabolism. Bethesda, *MD: Federation of the American Societies for Experimental Biology.*
29. Zakay D. and Block R.A., 1997, Temporal cognition. *Current Directions in Psychological Science*, 6, 112-116.
30. Church R.M., 1989, Theories of timing behavior, *Contemporary learning theories: Instrumental conditioning theory and the impact of biological constraints on learning*, Klein and Mowrer (eds.), Hillsdale, NJ: Erlbaum, 41-71.
31. Binkofski F. and Block R.A., 1996, Accelerated time after left frontal cortex lesion, *Neurocase*, **2** 485-493.
32. Davies P., 1994, *The last three minutes*, Basic Books, New York.
33. Hindmarch I., Quinlan P.T., Moore K.L., and Parkin C., 1998, The effects of black tea and other beverages on aspects of cognition and psychomotor performance, *Psychopharmacology*, **139** 230-238.
34. Means L.W., Higgins J.L., and Fernandez T.J., 1993, Mid-life onset of dietary restriction extends life and prolongs cognitive functioning, *Physiology and Behavior*, **54** 503-508.

Life Experience and the psychological suffering: psychiatric patients' perception

MARIA ALICE PEREIRA, MARIA CECÍLIA MORAES SCATENA, AND RENATA CURI LABATE
Dept. of Psychiatric Nursing & Human Sciences, Nursing School of Ribeirão Preto, University of São Paulo, Brasil.

1. INTRODUCTION

The concepts and views about mental illness have changed throughout history. Many trends and different schools of thought have been influencing the evolution of concepts about mental illness, which is also represented in various fields of knowledge.

In the field of Arts, such an illness is present in painting, sculpture, theater and literature. In the tale "O Alienista" (The alienist), by Machado de Assis[1], there is a character who searches for its essence. However, the essence is not revealed and the doubts about it and about who would actually be the mentally ill remain.

Mental illness is difficult to be treated. Studying it means to study the excluded, alienated, unwelcome people, that is, those regarded as not fitting in human culture and society.

Throughout history, each period approached mental illness differently. It is considered remarkable due to its staggering, unique, irreverent, multifaceted characteristic, and is also associated to psychological suffering. Hence the importance of focusing people enduring mental illness considering what they express in their life history, the meanings, behaviors, and not the disease itself, even because little is known about it.

For Basaglia[2] the history of mental illness is the history of a judgement, of gradual evolution of values, rules, systems of power under which are formed social groups and the organization of life. History of madness also

Studies on the Structure of Time: From Physics to Psycho(path)ology
Edited by Buccheri *et al.*, Kluwer Academic / Plenum Publishers, New York, 2000.

means a history of reasoning. Reasoning is identified with the power it shares out to some groups, some modalities, in specific times and places, in terms of its language and logic, and restrict on of general human needs.

Rules and values may suffocate or hinder the expression of needs and desires, leaving little room for expression of subjectivity in psychiatric patients.

The patients studied by Pereira[3] mentioned different etiological conceptions. They had elaborated theories about mental illness with particular characteristics. Each subject had his/her own theory for the symptom, according to their subjectivity and life experience. They also said that living with mental illness is living in the shadow of the outsiders' land, victims of loss and deprivation. None of the patients assumed the position of "mad", but of being in a mentally ill state. When the word "madness" was mentioned in the interviews, it was always attributed by others, not as a permanent state. The verb "estar" ("to be", in English) was used mostly in the past tense, suggesting that mental illness could mean a state of disease in a certain period of life:

"...I was sick in my mind", "I got mentally ill", "I cracked up".

The everyday life inside a traditional mental institution is characterized by intense psychological suffering and by the absence of tangible perspectives, leading the patients to helplessness and disbelief. We could not foresee possibilities of service improvement. Interventions that could encourage effective changes to better quality of service did not happen.

The background underlying the lack of perspectives for the mentally ill is the traditional psychiatric assistance model, that is, the medical model which brings and imposes the idea of an unchangeable situation for the "mentally ill".

Rotelli[4] says that psychological suffering tends to be disregarded in psychiatry, hence its transformation in disease. He criticizes the pseudo-scientific, cultural, institutional and economic apparatuses to take care of the mentally ill, and says that such transformation reflects and operates an institutionalization of suffering according to private objectives.

Gadamer[5] says that the soul is not a part, but the very existential corporeal complexity of the man. He emphasizes that Aristotle knew that when he considered the soul as the body's life. As well as Plato, he claimed that the body could not be cured, if there is no attempt to cure the soul.

So, listening to the "differences" of mental illness, and taking into consideration the psychological suffering originated in the condition of being "mentally ill", are attempts to integrate, elaborate and maybe to transform this condition, since who endures such an illness is exposed to the influences of society and culture as a subject/object of a social suffering.

Our intention as we listened to the patients was to learn their representations about madness, internment and the suffering originated from the condition of "mentally ill". This paper, as part of a broader study, aims at identifying the representations of psychiatric in-patients about their lives after living with madness, internment and psychological suffering, considering their perceptions about life experiences.

2. THEORETICAL-METHODOLOGICAL PROCEDURE

This is research is based on qualitative methods of investigation.

For Kirk and Miller[6] these methods have been used in social sciences, regarding the observation of people in their own territory, interaction with these people in their own language, and trying to understand the nature of the object considering the context of the phenomena.

According to Kirk and Miller[6], qualitative research is pluralistic. It allows the use of many techniques, provides a broader perspective and deeper understanding of the phenomenon.

2.1 The research context

The Psychiatric Hospital of Ribeirão Preto was chosen as the field of research due to its partnership with the public health system, and to the great number of patients from the city of Ribeirão Preto, as well as from surrounding cities. (There is another psychiatric hospital in the city, that is philanthropic with private partnerships).

2.2 The subjects

Twelve patients were selected according to typicality, that is, patients with history of life in the institution, or internment experience, able to tell their experiences during the years in the institution. Therefore, they were all "institutional" patients.

Typicality is directly related to "psychiatric career", that is, the pathway ran by the mental ill[7].

2.3 Life History

The Life History technique originated from American anthropology and psychology between the 1920's and 30's[8]. Sociology and other fields of knowledge have adopted it since then.

For Queiroz[8] Life History is "a technique in the intersection of what is external to the individuals and what they have in privacy, between the social and individual".

Another study by Queiroz[8] emphasizes the emergence of spontaneous narration as closely related to free associations in psychoanalysis. Bastide[9] mentioned the freedom such association provides, calling it "Proust's technique": "The interviewee, neglecting the astronomer's chronology, let go of him/herself as the young boy Proust to the "intermittence of the heart", infinitely more revealing then the former, under the apparent disorder of dates".

Therefore, listening to psychiatric patient's individual histories allowed us to enter an undiscovered world. The direct contact with these people *in loco* was very important, allowing the approach to their individual reality.

2.4 Collecting the data and developing the technique

The observation technique consists of direct contact of the researcher with the phenomenon to be observed, aiming to obtain information about the reality of the social actors in their own context[10]. In this process, the observer, as part of the observation context, directly interacts with the people observed.

Open, non-directive interviews were used to collect data, since they allow more flexibility and broader investigation about the interviewee[11].

Interviews of approximately one-and-a-half hour were recorded with the permission of the subjects. All of them were invited to talk about their lives with the following statement:

"We are writing a book about the people who are in a psychiatric hospital, and we would like to know if you can collaborate with us, talking about your life".

After they agreed, the interviews were guided by the subject's free associations.

2.5 Data analysis

Our first approach to with the transcriptions was inspired by Silva[12], who emphasizes the importance of allowing the data to rest for some time in the researcher's mind, bearing overwhelming stimuli and absence of meaning. For Silva[12], caution is important at this moment, when "the unconscious does its job and the meanings finally emerge, relieving the researcher from the angst and blessing him with provisory theorizing, originated from his/her study material".

Next, we used a thematic analysis procedure, from the first phase of *Statement analysis* by D'Urung[13]. D'Urung elaborated this technique aiming at a qualitative treatment of data obtained in non-structured situations. It is a modality of *Content analysis* and differs from other techniques in 2 important characteristics: 1) it is based on a conception of communication as a process, not as data; 2) it deviates from structural and formal analisys. This technique is adequate to the type of discourse produced by the non-directive interview.

The internment environment and the position of each subject form the context in which the ideas and conceptions are developed. The patients elaborated theories that built up to social representations. Jodelet[14] understands social representations as "a form of knowledge shared and elaborated socially, with a practical finality, and contributing to the construction of a common reality for the society as a whole."

Bleger's conjectures of "human conduct"[15] are also adequate for this study. For Bleger, human conduct is an effect or response from a given situation, and its manifestations co-exist in the body, mind and outside world.

2.6 Results and discussion

The material obtained was organized in categories, presenting the subjects' perceptions, experienced individually within the social context.

2.6.1 Internment as a period of exclusion, segregation, alienation, loss of freedom, punishment and absence of rights

The discourses imply to the hospital the meanings of prison and confinement, emphasizing the absence of freedom when the patients realize they cannot decide anything about their lives. During internment, they are not allowed to make choices even if regarding simple things in the hospital routine. For instance, walk freely in the facilities, choose among some activities:

"I feel imprisoned, I haven't got any optional activity to spend time..."

According to Lingiardi and Maffei[16], the current assistance system favors the absence of feelings, and emptiness takes over. The appropriation of the body by the institution may lead to a loss of the sense of life, transforming the subject into an inanimate body in his/her suffering.

In some narratives, the hospital is even defined as a prison, a place for criminals who must be punished. This association may be possible due to the hospital's resemblance to a prison, since it uses a restrictive apparatus (tall walls, fences, and locked gates and doors).

The narratives present a clear perception of reality, and within such an awareness indignation emerges, since the patients do not feel themselves as criminals deserving it.

"This is not a hospital, it's a prison. I'm not a criminal to be in jail..."

"To take my medicine I could stay at home, with freedom, going wherever I want. This is a prison..."

The activities appear in narratives as a way of "filling in time". The activities are regarded as helping to put up with the internment period.

"It would be necessary the occupational therapy now, which is not happening. Making nets, rugs, it is not happening. So we're very inactive, that's what I think... I haven't got any optional activity to fill in the time..."

The experience of time is usually related to the kind of activity people do. Some interesting, pleasant activity causes a perception of time passing quickly, whereas doing nothing cause a sensation of slowness.

In the hospital, the routine is endlessly repeated, and inactivity is so intense that may lead to a disturbance of daily and weekly cycles. As "all days are the same", it becomes difficult for the patients to locate themselves in time, also because there are no calendars or clocks at the yards. Under such conditions, time disorientation does not happen because of the illness condition, but also by the inactivity predominating in the hospital and yet by the absence of time markers and references.

The subjects in this study showed that time passed always the same way, without unexpected or thrilling facts about special dates, months and years:

"Now comes Christmas next Monday, then New Year's eve, Carnival, then Good Friday, that's it..."

Basaglia[2] says that the mentally ill suffers above all for being forced to live with no problematic matters or demanding activities, in the institutional space. According to the current model of assistance, this space is always restrictive and limiting for the people living there, causing different effects and conducts. The subjects mention hearing screams, fears, weeps and moaning. Such narratives are related to uneven relationships, resembling torture procedures:

"They tied me up, I took lots of electroshocks, it was sad...I was afraid...I still weep bitterly..."

2.6.2 The "getting used to" it

The testimonies convey the idea that the in-patients "got used to" not expressing their pains and desires. Such a fact corresponds to the idea of passivity, waiting a time of hopelessness. Below are some testimonies where the "getting used to" is present, where we noticed a contradiction, which leads our thoughts to the question: "Could somebody "get used to" pain?"

"Yeah, I got used to stay here. I didn't want to stay here, but now I got used to it..."

"I was inconsolable, and wept and wept...nowadays I'm more resigned about the things".

According to Goffman[7], humiliation of the self leads to radical changes in the moral domain. Such changes could affect the subjects' beliefs about themselves, and about their beloved or significant others.

The break up between internment and the outside world lasts a long time, causing the rupture of former roles, losses, desolation, feelings of being guilty and sadness.

"I feel needy, I cry... at first I only wept, I feel needy, I feel lonely, resentful of the things that happened in my life".

Thus, the psychological suffering remains throughout the subjects' lives as they tell their histories. The patients showed that thinking about the past, about the facts gone, brings bitter feelings. Thinking about the present and living it, however, also makes them fear the future:

"...When I got ill, I had a lot of thought, and then I got ill of so much thinking, too much thinking...I still get very worried about it..."

"...I'm afraid of some day becoming ill like that patient."

2.6.3 Mental illness as an adjective of time

In the narratives, mental illness appears as an event, a "becoming ill" in the process of existence.

"...I've been ill for some time..."

"...I became mentally ill"

"...I was ill"

We may think that the mentally ill do not consider mental illness as an absolute state. The impairment was not mentioned as a substantive, the verbs "to be" and "to become" are highlighted in the past tense. Therefore, the current condition of mentally ill, institutionalized in-patient, experienced by each one, is considered as an adjective, a condition due to factors related to the social condition and current assistance model.

We also noticed the idea of surveillance to the experience in the institution so far, resisting throughout the path, as in the narrative below:

"Some people even died of electroshock! Thank God I survived and here I am, safe and sound..."

The word "pain" is hardly mentioned, what does not happen with the word "suffering". The suffering is mentioned impersonally, characterized by bitter facts, having different meaning for each subject.

Although these people have mentioned the mental distress as an adjective, life is remembered with sorrow and sadness.

"My sad life, I'm very tired"

"...I lived my life like this, suffering and needy in the world."

"...A person of a hard life, suffered a lot inside, but she doesn't show on her face"

2.6.4 The time passed impassively

Time passed impassively, as recalled in the subjects' narratives. Sometimes the narrative was nostalgic, but always with the implacable sign of bitterness:

"In my first internment here I was pretty young, I was still pretty young"

"The old times were hard, I suffered too much..."

"...I came here a long time ago..."

Regarding these narratives, our analysis was supported by Comba[17], for whom the mental institution is an island left aside by society. This "island" would be equipped and used as a "last stop". "This place is an unchangeable world, with no history, into where time does not cross the gates. Inside, the days pass by indistinctly, identical and empty. To count time, one would have to draw a cross daily on the wall."

We had narratives full of memories, which presented the last hours before entering the hospital as a "near past", even when they happened decades ago.

Facts regarded as important, inside the hospital, were narrated as happing day before, or years ago.

The years passing by led to the idealization of another time not experienced, left behind, which could have been different, as if the past calendars were "badly programmed".

"...I suffered a lot, I saw many things..."

"...I'm going towards old age, I'm 63 years old...I'm getting heavier..."

The grieves along this way left little room for freedom or possibilities of leading roles in their own lives.

Although the subjects talked about the present, they conveyed the idea that the pains are stored in the past. One of the patients talked about herself as "Joana suffered from the past".

Remembering the years passing by brings the feeling that these are lost moments in existence, destroyed or taken away from the person. Those would be worthless times or memories, and should be extinct.

"I think I didn't have to live in a hospital like this. I think there should be someway else..."

The subjects presented the feeling of time stagnation, by showing resignation, abnegation of desires as a way to survive, lessen the pain, concluding that nonconformism would lead to more pain.

One of the interviewees said he feels sorry for the patients who have not learned to resign:

"The patient told me – 'I want to get away from here! Won't anybody pick me up?' – I felt sorry for her...for the time I used to cry a lot for everything, it takes longer to heal and pass."

3. CONCLUSIONS

This work aims to identify how institutionalized psychiatric patients perceive time, considered as an important factor to understand shared reality.

In the narratives of those living with a mental illness, we could see that as they talk about their lives in the present, they mentioned also the past. At the same time, they construct a "time network", drifting among dimensions in the time lived, organizing the experiences throughout life and representing time separately. In this way, temporality is presented as one of the dimensions of being, and the experiences in time happen along each person's development.

The interviewees showed they see the hospital as a world where their freedom is ceased, with inactivity, aggressive situations, losses, loneliness, confinement and imprisonment. It is not seen as a therapeutic place, but on the contrary, our subjects showed that the long internment period should be erased, as a time they did not lead their own lives.

By following the life histories in the narratives, we may think that from the moment of "going ill", culminating with the internment, these patients did not have any changes bringing new meanings to their lives, nor improving their condition. They are presented as victims of a health system, with a faulty assistance, and are fully aware of how harmed they are by this very system. This indicates they are at the same point where they started their "psychiatric career", although some have been in-patients for 15 to 50 years, and others have had many entries in the institution.

Nevertheless, we can see that the subjects listen to so many voices, which are lively contradictions, showing traces of life. These contradictions, active and vibrating, or sometimes enclosed in the narratives' subjacent meanings, show energies, polarities, ignored or numb potentialities, retained in personal resources.

REFERENCES

1. Assis M., 1962, O alienista, *Machado de Assis Obra Completa, v. 2*, José Aquilar, Rio de Janeiro.
2. Basaglia F., 1982, Ragione/Follia, Bisogno/Follia, *Basaglia Scritti, v. 2*, Einaudi, Torino.
3. Pereira M.A., 1997, *Representações Sociais de Pacientes Psiquiátricas sobre a loucura, a internação e o sofrer psíquico: a triste passagem e a triste paisagem, Ribeirão Preto,* 198 Tese (doutorado), Escola de Enfermagem de Ribeirão Preto, Universidade de São Paulo.
4. Rotelli F., 1993, *Per la normalità: taccuino di uno psichiatra*, Trieste Edizioni.
5. Gadamer H.G., 1994, Ermeneutica e psichiatria, *Dove si nasconde la salute*, Raffaello Cortina, Milano.
6. Kirk J. and Miller M.L., 1986, *Reliabily and validity in qualitative research, 2 ed,* California, Sage.
7. Goffman E., 1990, *Manicômios, prisões e conventos. 3 ed*, Perspectiva., São Paulo.
8. Queiroz M.I.P., 1988, Relatos orais: do indizível ao dizível, *Experimentos com histórias de vida (Itália-Brasil)*, Vértice, 14-43, São Paulo.
9. Bastide R., 1953, Introdução a dois estudos sobre a técnica das histórias de vida, *Sociologia, v.15*, 1 3-7.
10. Cruz Neto O., 1994, O trabalho de campo como descoberta e criação, *Pesquisa social: teoria, método e criatividade*, Vozes, Rio de Janeiro.
11. Bleger J., 1993, *Temas de psicologia, Entrevistas e grupos*, Trad. de Rita Maria M. de Moraes, Martins Fontes, São Paulo.
12. Silva M.E.L., 1993, Pensar em psicanálise, *Investigação em psicanálise*, Campinas.
13. Bardin L., 1994, *Análise de conteúdo*, Edições 70, Lisboa.
14. Jodelet D., 1989, *Folies et représentation sociales*, Presses Universitaires de France, Paris.
15. Bleger J., 1989, *Psicologia da Conduta*, Artes Médicas, Porto Alegre.
16. Lingiardi V. and Maffei C., 1994, Capire la sofferenza mentale, *KOS-rivista de medicina, cultura e scienze umane*, **109** 28-31.
17. Comba L.J.C., 1985, Mulheres: o último setor fechado, *A instituição negada: relato de um hospital psiquiátrico. 2 ed*, Basaglia (coord.), Graal, Rio de Janeiro.

Time in biological systems

JONATHAN D.H. SMITH
Department of Mathematics, Iowa State University, Ames, IA 50011, USA

1. INTRODUCTION

Biology is the keystone discipline in the bridge between the reversible time of microscopic physics and the diverse temporal phenomena of psychological experience. Biological systems are characterized by a range of time parameters that differ qualitatively and quantitatively from the underlying physical time. This paper will focus on two important examples of biological times: a macroscopic, statistical time that emerges from a simple model of species competing for limited resources, and a logarithmic time scale governing the aging process in humans. Each of these main examples is presented at two levels: in summary verbal form for readers who prefer to avoid mathematical details, and then more precisely using mathematical language. Some concluding speculations examine the potential significance of logarithmic time scales for the process of ontogeny.

2. COMPETITION BETWEEN SPECIES

To illustrate the competition model, consider a large lake offering a supply of nutrients constantly maintained at a level that will support a fixed number of fish. The fish may belong to one of three different species. In the absence of competition, and in the presence of an unlimited food supply, each of the species has its own natural growth rate. These growth rates differ between the species: describe them as small, medium, and large. When competing together for the fixed food supply in the lake, the fish population undergoes gradual changes. Although initially the proportions of the three

Studies on the Structure of Time: From Physics to Psycho(path)ology
Edited by Buccheri *et al.*, Kluwer Academic / Plenum Publishers, New York, 2000.

species may be roughly equal, the most rapidly reproducing species eventually comes to dominate. This natural dominance is irreversible within the system. The age of the population could be determined by a fisherman making a sample catch of the fish. The larger the proportion of the most prolific species within the catch, the older the population. Mathematically, the system is described by the so-called Gibbs canonical ensemble[1], the same model used to describe the distribution of particles amongst the various energy states of a physical system in thermal equilibrium. The natural growth rates of the fish species correspond exactly to the energies of the physical states. Most importantly, the age of the fish population corresponds exactly to the (inverse) temperature of the physical system. The temperature of, say, the water in a lake is a statistical quantity determined at a scale above the microscopic. Similarly, the biological time parameter measuring the age of the fish population is a statistical property of the population. The fisherman's catch determines the age of the population in the same way that a thermometer measures temperature. One could rejuvenate the fish population by restocking the lake with the less prolific species, just as one could change the temperature of the lake by admitting a flow of warmer water to it. Thus the time parameter giving the age of the population is characterized by three qualitative differences from local physical time. The age is statistical in nature, irreversible within the system, and reversible from outside the system.

3. ENTROPY, PROBABILITY, AND RANDOMNESS

For the scientific analysis of complex systems, the three concepts of entropy, probability, and randomness are fundamental. Nevertheless, they have proved to be so elusive, with such a plethora of subtle ramifications, that their use has often tended to lead to considerable confusion[2]. This section, which is a preliminary to the precise treatments of the biological times discussed in the paper, is an attempt at distilling current mathematical views of the concepts into readily accessible working definitions.

Consider a scientific experiment which may have any one of N possible outcomes. For example, N=6 for the experiment of rolling a die. By convention of modern science, a "scientific" experiment has to be repeatable and reproducible. The experiment is said to be *random* if no statistical test available to the experimenter can detect any pattern in repeated outcomes. It is very important to note that this definition is contingent on the power of the apparatus available. Rolling a die under casino conditions should be random. On the other hand, if equipment such as a high speed precision camera was available, it would be possible to predict the outcome of each roll from the initial motion of the die, and the experiment would no longer be random. A subtler distinction arises when rolling a very slightly loaded die under casino conditions. If the number of rolls required for even a sophisticated statistical test to reveal the bias is in excess of the number of rolls sufficient to wear the spots off the die, then the experiment would still qualify as random. The basic concepts of entropy and probability follow from the concept of randomness. For a random experiment with N outcomes, the (*natural*) *entropy* is

$$H = \log N \tag{3.1}$$

(using natural logarithms) or the (*binary*) entropy

$$H = \log_2 N \text{ bits} \tag{3.2}$$

using logarithms to base 2 and bits as units. (Occasionally logarithms to base 10 are used, with Hartleys as units.) The *probability* $\pi(x)$ of any one particular outcome x of the random experiment is

$$\pi(x) = N^{-1}. \tag{3.3}$$

The probability $\pi(x)$ and natural entropy H are connected by the mutually inverse relationships

$$\begin{cases} (a) \quad H = -\log \pi(x); \\ (b) \quad \pi(\mathrm{x}) = \exp(-\mathrm{H}). \end{cases} \tag{3.4}$$

Using binary entropy H, these take the form

$$\begin{cases} (a) \quad H = -\log_2 \pi(x); \\ (b) \quad \pi(x) = 2^{-H}. \end{cases} \tag{3.5}$$

The probability $\pi(x)$ represents a fair stake to buy into the following game: win one unit if the outcome of the experiment is x. Randomness of the experiment means that there are no winning strategies in this game. The entropies (3.1) and (3.2) measure one's ignorance about the outcome of the experiment. If you let someone else run the experiment, and instead question them afterwards as to what the outcome was, then the minimum number of yes/no questions required to elicit the outcome would be given by (3.2).

The above formulae are too narrow to be of general use, where one wishes to deal with non-random experiments. These may be modeled using an underlying random experiment whose set of outcomes, called the *phase space*, has N elements. The phase space is completely partitioned into a set $\xi=\{C_1, \ldots, C_r\}$ of mutually exclusive subsets called *states*. The partition ξ represents the non-random experiment (also denoted ξ) of sampling an outcome x from the phase space and locating the state C_i in which it lies. If the state C_i contains n_i outcomes of the underlying random experiment, each of whose outcomes has probability N^{-1} according to (3.3), then the *probability* $p(C_i)$ of the state C_i is given as

$$p(C_i) = n_i N^{-1}. \tag{3.6}$$

The state C_i may be regarded as a random experiment in its own right: select an outcome from C_i. The entropy of this random experiment, according to (3.1), is

$$H(C_i) = \log n_i. \tag{3.7}$$

If you perform experiment ξ and obtain the result C_i, then your ignorance will have been reduced by $\log N - \log n_i = -\log p(C_i)$. This happens with probability $p(C_i)$. Thus the average loss of ignorance or gain in knowledge obtained on performing experiment ξ is its *entropy*

$$H(\xi) = -\sum_{i=1}^{r} p(C_i) \log p(C_i). \tag{3.8}$$

(Of course, one may take logarithms to base 2 and quote $H(\xi)$ in bits.) The mathematical discipline of measure theory extends the definitions of probability and entropy to appropriate infinite phase spaces, where "counting outcomes" may be replaced by "measuring volumes". The entropy $H(\xi)$ satisfies the inequality

$$0 \leq H(\xi) \leq \log r. \tag{3.9}$$

Equality obtains on the left in (3.9) if and only if $p(C_i)=1$ for some i: if you already know in advance that ξ will come up with state C_i, then you gain no knowledge by performing the experiment. Equality obtains on the right in (3.9) if and only if $p(C_i)=r-1$ for each i: the most informative experiments are those designed so that all their different outcomes are equally likely. In particular, randomness of an experiment is characterized by its entropy. Moreover, the three concepts of "entropy", "probability", and "randomness" turn out to be equivalent. Each leads to the other two. In complexity theory, it is an entropy measure, namely the self-delimiting algorithmic complexity, which is usually taken as basic. Probability is then obtained via (3.5)[3]. The complexity class of the algorithms invoked corresponds to the power of the apparatus used in the statistical tests for randomness above.

4. STATISTICAL MECHANICS: THE CANONICAL ENSEMBLE

For the experiment $\xi=\{C_1, \ldots, C_r\}$ considered in the previous section, absolute randomness - complete ignorance about the outcome - was characterized by unconstrained maximization of the entropy $H(\xi)$, attaining the value $\log r$ according to (3.9). In practice it may be possible to assign a numerical value E_i to each state C_i, e.g. the number of spots on the face of a die or an energy in electron volts. If the expected value

$$E = \sum_{i=1}^{r} p(C_i)E_i \tag{4.1}$$

is known, then the non-negative probabilities $p(C_i)$, which have to satisfy the relationship

$$1 = \sum_{i=1}^{r} p(C_i), \tag{4.2}$$

are determined by the assumption of relative randomness: no pattern is discernible in repeated outcomes except for maintenance of the fixed value E. This is equivalent to maximization of the entropy $H(\xi)$ subject to the constraints (4.1) and (4.2). Setting

$$g_1 = E - \sum_{i=1}^{r} p(C_i)E_i, \quad g_2 = 1 - \sum_{i=1}^{r} p(C_i)$$

and

$$f = H(\xi) + \beta g_1 + \lambda g_2$$

with Lagrange multipliers β, λ, the stationarity conditions $\partial f/\partial p(C_i)=0$ lead to $\log p(C_i) = -\beta E_i - (1+\lambda)$ or $p(C_i) = \exp(-\beta E_i)/\exp(1+\lambda)$. Substituting into (4.2), noting that λ is independent of i, one obtains

$$p(C_i) = Z(\beta)^{-1}\exp(-\beta E_i) \tag{4.3}$$

with the *partition function* or *Zustandssumme*

$$Z(\beta) = \sum_{i=1}^{r} \exp(-\beta E_i). \tag{4.4}$$

The fixed value E from (4.1) is recovered as

$$E = -\frac{d\log Z(\beta)}{d\beta}; \tag{4.5}$$

on the other hand this equation may yield β. The entropy (3.8) may be recovered via

$$-\sum_{i=1}^{r} p(C_i)\log p(C_i) = \sum_{i=1}^{r} p(C_i)[\beta E_i + (1+\lambda)]$$

as

$$H(\xi) = \beta E + \log Z(\beta). \tag{4.6}$$

Note that (4.3)--(4.6) are just consequences of the assumption of randomness of ξ relative to (4.1) and the attribution of the numerical value E_i to each state C_i. There is no inherent reason for such an experiment ξ to be inappropriate as a model in certain biological circumstances. (In many cases, as in §8 below, additional numbers F_i, G_i, etc. may be assigned to each state, with known expected values F, G, etc. The Lagrange multiplier method readily extends to such cases, using vectors in place of scalars.) An experiment ξ with random probabilities subject to (4.1) is called a *canonical ensemble*, since it generalizes the models of that name in (equilibrium) statistical mechanics. A correspondence with thermodynamics arises when the E_i and E are energies in suitable units. Introducing *Boltzmann's constant* K, the *thermodynamic entropy* is

$$S = KH. \tag{4.7}$$

The *temperature* is

$$\Theta = 1/K\beta. \tag{4.8}$$

The *thermodynamic potential* is

$$\Psi = \log Z(\beta). \tag{4.9}$$

The *Helmholtz free energy* is

$$F = -K\Theta\Psi. \tag{4.10}$$

Equation (4.6) then reduces to

$$F = E - \Theta S. \tag{4.11}$$

All this formalism is available in the general context of the canonical ensemble, although the appropriate units may vary from case to case.

5. THE CANONICAL MODEL OF COMPETITION

The application of the canonical ensemble (with non-thermodynamic parameters) to the study of competition between species arises through Eigen's phenomenological rate equation[4] for the case of constant overall organization and negligible mutability. Consider r species 1, ..., r. (The verbal discussion of §2 assumed r=3.) Suppose that the i-th species has a known net growth rate E_i. In other words, if a population of n_i units of the i-th species is allowed to develop without constraint, its rate of change is given by

$$\dot{n}_i = E_i n_i. \tag{5.1}$$

Note that the units of the quantity E_i in equation (5.1) are $(\text{time})^{-1}$. Now suppose that the r species are brought together in a joint population maintained at a constant count N (e.g. by control of a common food supply). Thus the birth of one individual has to be compensated by the death of another, not necessarily of the same species. If n_i now represents the number of individuals of species i present in the joint population, its rate of change is given by

$$\dot{n}_i = (E_i - E)n_i. \tag{5.2}$$

Summing the equations (5.2) over i, constancy of the joint population amounts to

$$0=\sum_{i=1}^{r}\dot{n}_i=-NE+\sum_{i=1}^{r}n_iE_i. \tag{5.3}$$

Consider the experiment ξ of capturing a random individual from the population and determining its species. State C_i corresponds to the i-th species. Equations (3.6) and (5.3) yield the constraint

$$E=\sum_{i=1}^{r}p_iE_i \tag{5.4}$$

on the experiment ξ. IF NOTHING ELSE IS ASSUMED TO BE KNOWN (including the initial values of the n_i), then the canonical ensemble is the appropriate model for ξ, and (4.3) determines the probability of C_i as

$$p_i=\exp(-\beta E_i)/Z(\beta) \tag{5.5}$$

with

$$Z(\beta)=\sum_{i=1}^{r}\exp(-\beta E_i). \tag{5.6}$$

To interpret this, set

$$t=-\beta. \tag{5.7}$$

Since the dimensions of E_i are $(\text{time})^{-1}$, the dimensions of the Lagrange multiplier t are those of time. Indeed, rewriting (5.5) and (5.6) in terms of t yields

$$p_i=\exp(E_it)\Big/\sum_{j=1}^{r}\exp(E_jt) \tag{5.8}$$

or

$$n_i=N\exp(E_it)\Big/\sum_{j=1}^{r}\exp(E_jt). \tag{5.9}$$

Thus the Lagrange multiplier (5.7) is the intrinsic, statistical time parameter giving the age of the population. One may contrast it with the physical time implicit in Newton's "dot" notation in equations (5.1) and (5.2). It is instructive to compare (5.9) with a conventional solution of (5.2) as a Bernouilli differential equation[4,5,6]. The conventional solution uses the Ansatz

$$n_i = Nx_i \Big/ \exp \int_0^t E(\tau)d\tau, \tag{5.10}$$

which leads to the linear equations

$$\dot{x}_i = E_i x_i. \tag{5.11}$$

In deriving (5.9), nothing was assumed to be known about the initial distribution of the n_i. The canonical ensemble model, maximizing the entropy $H(\xi)$, thus selects the maximum-entropy initial distribution

$$n_i(0) = N/r, \tag{5.12}$$

i.e. $x_i(0) = r^{-1}$. With this initial condition, the conventional solution of (5.2) is

$$n_i = N\exp(E_i t) \Big/ r\exp \int_0^t E(\tau)d\tau. \tag{5.13}$$

Comparison with (5.9) yields the partition function

$$Z(t) = \sum_{j=1}^{r} \exp(E_j t) = r\exp \int_0^t E(\tau)d\tau. \tag{5.14}$$

As a potential, the thermodynamic potential (4.9) is only determined to within an additive constant. Using (5.14), one may thus express the "thermodynamic potential" in the present context as the "energy integral"

$$\Psi = \int_0^t E(\tau)d\tau. \tag{5.15}$$

By (5.4), $E(t)$ is the cull rate necessary to hold the total population constant. Thus (5.15) represents the total number of individuals lost to population control over the first t time units. The "Helmholtz free energy" (4.10) becomes the average cull rate

$$F = \frac{1}{t}\int_0^t E(\tau)d\tau \tag{5.16}$$

over the first t time units.

6. HUMAN DEMOGRAPHY

The second biological time parameter to be considered arises in the demography of human populations. To simplify the analysis, one usually focuses exclusively on the female population (*pater semper incertus est*). The population is divided into successive age classes $\ldots, a_i, a_{i+1}, \ldots,$ typically taken to be five years in length. The chance of an individual's passage from one age class a_i to the next is described by a survival probability b_i specified as in an actuarial life table. While sojourning in a particular age class a_i, a female gives birth to a specified average number m_i of daughters. The so-called *net maternity function* assigns a numerical value

$$m_i \prod_{j<i} b_j \tag{6.1}$$

to each age class. This value is the probability that a new-born female will survive and enter the age class, multiplied by the average number of daughters born to females in that age class. In conjunction with the probability distribution for the age of the mother of a randomly chosen new-born, the net maternity function specifies all the significant intensive demographic variables, such as the generation time, family size, and overall growth rate of the population. As a function of age itself, the net maternity function displays a skewed pattern that does not lend itself to ready analysis (ref. 7, Chapter 6). But if one uses the logarithm of the mother's age as the independent variable for the net maternity function, then the function values rescale to give the familiar bell-shaped curve of the Gaussian or normal distribution. In other words, the key time parameter for the aging process described by the net maternity function is the logarithm of the calendar age. This accords with Backman's concept of *organic time*[8,9] and Prigogine's concept of *thermodynamic time*[9,10]. Similar effects have been noted at the psychological level: the perceived intensities of memories and experiences in the period from ten to twenty years of age roughly match those from twenty to forty years, and again those from forty to eighty years. Viewed logarithmically, these periods are of equal duration (doubling of calendar age over the period).

7. PROJECTION MATRICES AND LOTKA STABILITY

To begin the more detailed analysis of human demography, consider a population with age classes $a_{-k}, \ldots, a_0, a_1, \ldots, a_n, a_{n+1}, \ldots, a_{n+l}$, arranged in order of increasing ages. An individual of age a_i has a constant positive probability b_i of surviving to the next age a_{i+1}. No individual of age class a_{n+l} survives to a later age class. The only reproductive age classes are $a_l, \ldots, a_n$. During their sojourn in each such reproductive age class a_i, individuals give birth to a positive average number m_i of new-borns falling into age class a_{-k}. In breakdowns of female human populations[11], one might have k=1, n=9, l=7, with a_{-1} the 0-4 year age class, ..., a_1 the 10-14 year olds, ..., a_{15} the 80-84 year olds, and a_{16} those aged 85 or more.

The process is represented by a directed weighted graph or directed network, the "clock of ages". There is a node for each age class. From each node a_i to a_{i+1}, for $i=-k, \ldots, n+l-1$, there is a directed edge of weight b_i. From each node a_i to a_{-k}, for $i=1, \ldots, n$, there is a directed edge of weight m_i. Demographic significance resides in the subnetwork obtained by deleting the post-reproductive vertices $a_{n+1}, \ldots, a_{n+l}$ and the edges incident with them. The adjacency matrix of this subnetwork is the (transposed) Leslie or *projection matrix*

$$A=\left[\begin{array}{ccccc|ccccc}
0 & b_{-k} & \cdots & 0 & 0 & 0 & 0 & \cdots & 0 & 0 \\
0 & 0 & & 0 & 0 & 0 & 0 & & 0 & 0 \\
 & & & & & & & & & \\
0 & 0 & & 0 & b_{-1} & 0 & 0 & & 0 & 0 \\
0 & 0 & & 0 & 0 & b_0 & 0 & & 0 & 0 \\
\hline
m_1 & 0 & & 0 & 0 & 0 & b_1 & & 0 & 0 \\
m_2 & 0 & & 0 & 0 & 0 & 0 & & 0 & 0 \\
 & & & & & & & & & \\
m_{n-1} & 0 & & 0 & 0 & 0 & 0 & & 0 & b_{n-1} \\
m_n & 0 & & 0 & 0 & 0 & 0 & & 0 & 0
\end{array}\right] \tag{7.1}$$

For $n>1$, the clock of ages has cycles $a_{-k}a_{-k+1}\ldots a_1a_{-k}$, …, $a_{-k}a_{-k}\ldots a_na_{-k}$ of coprime lengths, so that the matrix A is primitive[9,12,13]. The Perron-Frobenius Theorem[14,15] then implies that the non-negative matrix A has a real, positive, dominant simple eigenvalue λ with a positive (row) eigenvector u. Simplicity of λ means that its eigenspace just consists of scalar multiples of u. Dominance of λ means that λ exceeds the absolute value of any other eigenvalue of A.

For an explicit determination of λ and u, it is convenient to introduce some notation. For $i\leq n$, define

$$l_i = \prod_{j<i} b_j, \tag{7.2}$$

the probability that a new-born survives to age a_i. The characteristic equation of A may then be written in the form[16]

$$\lambda^{k+1} = \sum_{i=1}^{n} \lambda^{-i} l_i m_i \tag{7.3}$$

and the components of the eigenvector $u = [u_{-k}\ldots u_{-1}u_0u_1\ldots u_n]$ may be taken as

$$u_i = \lambda_{-i} l_i \tag{7.4}$$

for $-k\leq i\leq n$ (ref. 16, §6).

Now suppose that there is a constant time difference z between the successive non-post-reproductive age classes a_{-k}, …, a_n. If the components of the row vector $x^t = [x^t_{-k}\ldots x^t_0 x^t_1\ldots x^t_n]$ record the number of individuals in each such age class at time t, the projection matrix (7.1) may be used to describe the development of the population over time via

$$x^{t+z} = x^t A. \tag{7.5}$$

The population is said to exhibit *Lotka stability* at time t if x^t lies in the eigenspace of λ. One then has

$$x_i^{t+z} = \lambda x_i^t$$

for $-k\leq i\leq n$, so that

$$\log x_i^{t+1} = z^{-1}\log\lambda + \log x_i^1. \tag{7.6}$$

The logarithmic increase

$$r_1 = z^{-1}\log\lambda \tag{7.7}$$

over unit time is called the *Lotka growth rate*.

8. THE CANONICAL MODEL OF MOTHERS' AGES

Now consider the experiment of choosing a new-born at random, and determining the age class of her mother. There are no assumptions of Lotka stability or constant difference z between the non-post-reproductive age classes. Suppose that the mother lies in age class a_i with probability q_i, for $1 \le i \le n$. One thus has

$$\sum_{i=1}^{n} q_i = 1. \tag{8.1}$$

Define the *generation time*

$$T = \sum_{i=1}^{n} q_i a_i \tag{8.2}$$

as the expected age of the mother. Define the (*logarithmic*) *maternity*

$$M = -\sum_{i=1}^{n} q_i \log l_i m_i \tag{8.3}$$

as the expected value of the negated logarithm of the net maternity function $l_i m_i$. Suppose that the numerical values of the generation time and maternity are known, but that one has no further information about the probability distribution $[q_1 q_2 \ldots q_n]$. As discussed in §4, the appropriate probability distribution for this model is the one that maximizes the entropy

$$H = -\sum_{i=1}^{n} q_i \log q_i \tag{8.4}$$

subject to the constraints (8.1), (8.2), (8.3). (Recall that the validity of a different distribution would amount to different knowledge of details of the experiment from that summarized by specification of the generation time and logarithmic maternity.)

Determination of the probability distribution is achieved using the Lagrangean function

$$L(q_i,\alpha,r,s) = -\sum_{i=1}^{n} q_i \log q_i + \\ +\alpha\left(1-\sum_{i=1}^{n} q_i\right)+r\left(T-\sum_{i=1}^{n} q_i a_i\right)+(1+s)\left(M+\sum_{i=1}^{n} q_i \log l_i m_i\right). \tag{8.5}$$

The stationarity conditions $\partial L/\partial q_i=0$ for $1\leq i\leq n$ lead to

$$\log q_i = -(1+\alpha)-ra_i+(1+s)\log l_i m_i \tag{8.6}$$

or

$$q_i = e^{-ra_i}(l_i m_i)^{1+s}/\exp(1+\alpha).$$

Substituting into (8.1), noting that α is independent of i, one obtains

$$q_i = Z(r,s)^{-1} e^{-ra_i}(l_i m_i)^{1+s} \tag{8.7}$$

with the partition function

$$Z(r,s) = \sum_{i=1}^{n} e^{-ra_i}(l_i m_i)^{1+s}. \tag{8.8}$$

The Lagrange multiplier r corresponding to the generation time constraint (8.2) is called the *Malthusian parameter*. The (shifted) Lagrange multiplier s corresponding to the maternity constraint (8.3) is called the *perturbation*.

9. COMPARISON WITH LOTKA-STABLE POPULATIONS

In order to understand the probability distribution (8.7) given by the canonical ensemble, and the interpretation of the Lagrange multipliers, it is useful to consider the case of a Lotka-stable population with a constant time difference z between successive non-post-reproductive age classes. In this case, (7.4) yields the probability

$$q_i = \lambda^{-i} l_i m_i \Big/ \sum_{j=1}^{n} \lambda^{-j} l_j m_j \tag{9.1}$$

for the mother of a randomly chosen new-born to lie in a reproductive age class a_i, namely $1 \le i \le n$. Note that

$$a_i = a_0 + iz \tag{9.2}$$

for $1 \le i \le n$. By (7.7), one then has

$$\lambda^{-i} = e^{-r_1 iz} = e^{r_1 a_0} e^{-r_1 a_i}. \tag{9.3}$$

The Lotka-stable probability (9.1) may thus be written in the form

$$q_i = Z(r_1, 0)^{-1} e^{-r_1 a_i} l_i m_i. \tag{9.4}$$

with

$$Z(r_1, 0) = \sum_{i=1}^{n} e^{-r_1 a_i} l_i m_i. \tag{9.5}$$

Comparing (9.4) and (9.5) with (8.7) and (8.8), it emerges that the Lotka-stable distribution is just the special case of the canonical ensemble distribution with zero perturbation, the Lotka growth rate being the Malthusian parameter of the canonical ensemble model. One may thus characterize Lotka stability within the canonical ensemble model as the freedom from perturbation.

10. MACROSCOPIC DEMOGRAPHY

The demographic analyses given by the projection matrix (7.1) or by the canonical distribution (8.7) have a major disadvantage: they involve large numbers of *microscopic parameters* such as b_i, l_i, m_i that have to be specified for each age class. Ideally, one should be able to give a *macroscopic* specification of a population, in terms of a small number of parameters that are independent of the age classes. To prepare for such a specification, it is convenient to use continuous variables in place of the discrete ones used hitherto. Thus one replaces the discrete probability q_i of §8 that the mother of a randomly chosen newborn lies in age class a_i by a probability density function $p(a)$ such that the probability of the randomly chosen newborn's mother's age A lying in the range $t_1 < A \leq t_2$ is

$$\int_{t_1}^{t_2} p(a)da. \tag{10.1}$$

The continuous net maternity function is denoted by $\varphi(a)$ (ref. 7, p. 100). The corresponding canonical density, the continuous analogue of (8.7), is

$$q(a) = Z(r,s)^{-1} e^{-ra} \varphi(a)^{1+s}. \tag{10.2}$$

Here the partition function is

$$Z(r,s) = \int_0^\infty e^{-ra} \varphi(a)^{1+s} da. \tag{10.3}$$

Setting s=0 into equation (10.3), one obtains

$$Z(r,0) = \sum_{k=0}^\infty \frac{(-r)^k}{k!} R_k \tag{10.4}$$

with

$$R_k = \int_0^\infty a^k \varphi(a)da \tag{10.5}$$

for natural numbers k (ref. 7, §6.1; ref. 17, §5.2). In particular, one has the *net reproduction rate* (ref. 7, p.102)

$$R_0 = Z(0,0). \tag{10.6}$$

Although the net maternity function $\varphi(a)$ is non-negative, it does not directly yield a probability density function on the set of ages. One might renormalize, considering

$$\varphi(a)/R_0 \tag{10.7}$$

as a density (ref. 17, §5.2). Attempts have been made to fit (10.7) with normal and other readily handled density functions on the set of ages (ref. 7, Chapter 6). However, as Keyfitz observes (ref. 7, p. 168), the attempts have not generally been satisfactory.

The essence of the current approach is to consider an appropriate renormalization of the net maternity function as a probability density function for the logarithm of the age, not for the age itself. The *renormalization constant* must be

$$\int_{a=0}^{a=\infty} \varphi(a)d(\log a)$$

or

$$R_{-1} = \int_0^\infty a^{-1}\varphi(a)da \tag{10.8}$$

[extending the notation (10.5)]. One then obtains

$$\varphi(a)/R_{-1} \tag{10.9}$$

as a probability density function. Under this density, the probability for the age random variable X to lie in a range $t_1<X\leq t_2$ is given as

$$\int_{a=t_1}^{a=t_2} \frac{\varphi(a)}{R_{-1}a}da. \tag{10.10}$$

One may also set $Y=\log X$, $y=\log a$ and use the equivalent expression

$$\int_{y=\log t_1}^{y=\log t_2} R_{-1}^{-1}\varphi(e^y)dy. \tag{10.11}$$

The expected value of the logarithm of the age under the density (10.9) will be written in the form

$$\log t = \int_0^{\infty} \frac{\varphi(a)\log a}{R_{-1}a}\,da \tag{10.12}$$

for a characteristic age t called the *logarithmic generation time*. Generally, one has

$$t < T \tag{10.13}$$

(The mnemonic "little t less than big T" is sometimes helpful.) The variance of the logarithm of the age under the density (10.9) will be written in the form

$$u^2 = \int_0^{\infty} \frac{\varphi(a)[\log(a/t)]^2}{R_{-1}a}\,da \tag{10.14}$$

with a non-negative, dimensionless parameter u called the (*standard*) *deviation*. Thus the random variable Y–$\log t$ has mean 0 and standard deviation u. Assuming that one has no further information about the distribution of Y, one concludes that $Y=\log X$ is normally distributed with mean $\log t$ and variance u^2 (ref. 18, p. 629) Equivalently, the age random variable X has the two-parameter lognormal distribution

$$\Lambda(\log t, u). \tag{10.15}$$

In other words (ref. 19, p. 2), the net maternity function is given as

$$\varphi(a) = \frac{R_{-1}}{u\sqrt{2\pi}} \exp\left\{-\frac{1}{2u^2}\left(\log\frac{a}{t}\right)^2\right\} \tag{10.16}$$

for $a>0$. Note its dependence on the three parameters: the renormalization constant R_{-1}, the logarithmic generation time t, and the standard deviation u. Making further use of the Malthusian parameter r and perturbation s, one may insert (10.16) into (10.2) to obtain

$$q(a) = \exp\left\{-ra - \frac{1+s}{2u^2}\left(\log\frac{a}{t}\right)^2\right\} \bigg/ \int \exp\left\{-ra - \frac{1+s}{2u^2}\left(\log\frac{a}{t}\right)^2\right\} da \tag{10.17}$$

or

$$q(a) = Z(r,s)^{-1}\left(\frac{R_{-1}}{u\sqrt{2\pi}}\right)^{1+s} \exp\left\{-ra - \frac{1+s}{2u^2}\left(\log\frac{a}{t}\right)^2\right\} \tag{10.18}$$

with

$$Z(r,s) = \left(\frac{R_{-1}}{u\sqrt{2\pi}}\right)^{1+s} \int_0^\infty \exp\left\{-ra - \frac{1+s}{2u^2}\left(\log\frac{a}{t}\right)^2\right\} da. \tag{10.19}$$

Together, (10.16) and (10.17) specify $\varphi(a)$ and $q(a)$ entirely in terms of the vector

$$(R_{-1}, r, s, t, u) \tag{10.20}$$

of macroscopic parameters. More succinctly, one may describe the shapes of $\varphi(a)$ and $q(a)$ by the proportionalities

$$\varphi(a) \propto \exp\left\{-\frac{1}{2u^2}\left(\log\frac{a}{t}\right)^2\right\} \tag{10.21}$$

and

$$q(a) \propto \exp\left\{-ra - \frac{1+s}{2u^2}\left(\log\frac{a}{t}\right)^2\right\}. \tag{10.22}$$

This macroscopic description of the intensive state of the population, known as the *canonical/lognormal model*, may then be summarized by (10.8), (10.21), and (10.22).

11. A SAMPLE POPULATION

As an illustration of the macroscopic specification of a population, consider the data for peninsular Malaysia in 1985 (ref. 11, pp. 380-1). The true counts of births by mother's age are presented in column 3 of Table 11.1 below. The true net maternity function is presented in column 5 of Table 11.1. The macroscopic parameters are R_{-1}=0.0642 per annum, r=0.0343 per annum, s=0.0733 (dimensionless), t=27.81 years, and u=0.2150 (dimensionless). Using these values, the canonical/lognormal model supplies the estimated births by mother's age listed in column 4 of Table 11.1, and the estimated net maternity function values listed in column 6 of Table 11.1.

Table 11.1. Malaysia 1985

1. Age Class	2. Age Group	3. Births by mother's age	4. Estimated births	5. Net Maternity function	6. Estimated nmf.
a_1	10-14	130	100	.0004	.0006
a_2	15-19	17,669	17,800	.0610	.0585
a_3	20-24	103,486	104,000	.3713	.3657
a_4	25-29	137,470	137,000	.5620	.5933
a_5	30-34	91,283	91,200	.4525	.4570
a_6	35-39	44,009	44,100	.2699	.2263
a_7	40-44	11,465	12,000	.0941	.0851
a_8	45-49	1,219	950	.0104	.0269
a_9	50-54	75	50	.0008	.0076

The agreement between the actual values and those given by the model is remarkably close, particularly in the demographically significant age classes.

12. ONTOGENY AND LOGARITHMIC TIME

The analysis presented in the preceding sections has shown that it is the logarithm of a mother's age, and not her direct calendar age, that is the most fundamental for a macroscopic description of the intensive demographic variables. Quite apart from reference to Backman[8] and Prigogine[9], one might ask why this logarithmic time should play such a role. One possible explanation may be found in the process of ontogeny of complex systems.

In hierarchical dynamical systems, low levels of the hierarchy function on rapid time scales, while high levels function on slower time scales. As a simple example of a two-level hierarchical dynamical system, consider the Bénard convection system in which a fluid in a confined space is heated from below[20]. The lower level of the hierarchy is represented by the random thermal oscillations of the molecules of the fluid. The characteristic time scale at this level is given by the mean time τ between intermolecular collisions (ref. 21, § 87). This time is of the order of 10^{-10} seconds. The higher level of the hierarchy is represented by the macroscopic motion of the fluid around the convection cells. The characteristic time of the higher level is of the order of seconds.

The derivative

$$\frac{d\log x}{dx} = \frac{1}{x} \tag{12.1}$$

of the logarithm function is large for small values of its (positive) argument, and small for large values of the argument. One would thus expect a logarithmic time to govern the process of ontogeny. Early in the process, low levels of the hierarchy are emerging at their characteristic rapid rates, so that the intrinsic time of the process moves forward quickly relative to physical time. Later on, the slower, higher levels of the hierarchy are being constructed. The intrinsic time at these stages moves less rapidly with respect to physical time.

REFERENCES

1. Smith J.D.H., 1996, Competition and the canonical ensemble, *Math. Biosci.*, **133** 69-83.
2. Hoffmeyer J., *Signs of Meaning in the Universe*, Indiana University Press, Bloomington, IN, (translation by Haveland B.J. of "En Snegl på Vejen: Betydningens Naturhistorie", Rosinante, København, 1993.).
3. Uspenskii V.A., Semenov A.L., and Shen' A.Kh., 1990, Can an individual sequence of zeros and ones be random?, *Russian Mathematical Surveys*, **45** 121-189.
4. Eigen M., 1971, Self-organization of matter and the evolution of biological macromolecules, *Naturwissenschaften*, **58** 465-523.
5. Jones B.L., Enns R.H., and Rangnekar S.S., 1976, On the theory of selection of coupled macromolecular systems, *Bull. Math. Biol.*, **38** 15-28.
6. Thompson C.J. and McBride J.L., 1974, On Eigen's theory of the self-organization of matter and the evolution of biological macromolecules, *Math. Biosci.*, **21** 127-142.
7. Keyfitz N., 1968, *Introduction to the Mathematics of Population*, Addison-Wesley, Reading, MA.
8. Backman G., Lebensdauer, and Entwicklung, 1940, *Arch. für Entwicklungsmechanik*, **140** 90-123.
9. Prigogine I., 1947, *Etude Thermodynamique des Phénomènes Irréversibles*, Dunod, Paris.
10. Bertelanffy L. von, 1968, *General System Theory*, Braziller, New York.
11. Keyfitz N. and Flieger W., 1990, *World Population Growth and Aging*, University of Chicago, Chicago, IL.
12. Demetrius L., 1971, Primitivity conditions for growth matrices, *Math. Biosci.*, **12** 53-58.
13. Rosenblatt D., 1957, On the graphs and asymptotic forms of finite Boolean relation matrices and stochastic matrices, *Naval Res. Logist. Quart.*, **4** 151-167.
14. Birkhoff G., 1967, Lattice Theory, *American Mathematical Society*, Providence, RI.
15. Gantmacher F.R., 1986, *Matrizentheorie*, Deutscher Verlag der Wissenschaften, Berlin.
16. Lewis E.G., 1942, On the generation and growth of a population, *Sankhya*, **6** 93-96.
17. Keyfitz N., 1985, *Applied Mathematical Demography* (2nd ed.), Springer, New York.
18. Shannon C.E., 1948, A mathematical theory of communication, *Bell System Tech. J.*, **27** 623-656.
19. Crow E.L. and Shimizu K., 1988, *Lognormal Distributions*, Marcel Dekker, New York.
20. Libchaber A. and Maurer J., Une expérience de Rayleigh-Bénard de géometrie reduite; multiplication, accrochage et démultiplication de fréquences, *J. Phys.(Paris) Coll*, **41** C3-51.
21. Rumer Yu. B. and Ryvkin M. Sh., 1980, *Thermodynamics, Statistical Physics, and Kinetics*, Mir, Moscow.

22. Parlett B., 1970, Ergodic properties of population I: the one sex model, *Theor. Pop. Biol.*, 1 191-202.

Temporal Aspects of Neuronal Binding

ALFREDO PEREIRA JR. AND ARMANDO ROCHA
Universidade Estadual Paulista, Brasil

1. INTRODUCTION

The construction of a unified perception of the world by the brain is one of the most intriguing problems studied by contemporary neuroscience. The physical environment provides perceiving organisms with a variety of stimuli, each of them based on different physical processes[1], although ultimately they could be described in terms of quantum "waves" of different lengths. For example, vision is based on fast and small particles, traveling in straight lines, while olfaction is based on slower and bigger particles traveling in brownian motion. Despite the physical differences of stimuli, the brain works in a way that visual images and odors are smoothly integrated into a unified perception[2]. Within each sensory modality, an integrative mechanism also works to bind different aspects of stimuli processed by specialized parts of the brain. Such a mechanism is plausibly a product of biological evolution, providing means for organisms to sense different parts of the environment and to support adaptive behaviors.

Binding phenomena are omnipresent in biological systems: proteins bind to effectors, organisms bind for reproduction, and in human language verbs bind to predicates. In cognitive neuroscience, a similar "binding" problem appeared, the problem of explaining how stimuli aspects within one modality and sensations from different modalities bind to produce a unified perception of the world[3,4]. Such a formulation conflates three inter-related but different problems:

a) the problem of *cortical* integration. Phylogenetic development of the neocortex led to the formation of specialized areas, called "primary sensory areas". Despite the existence of a large number of cortico-cortical connections, the mechanism that integrates the product of

Studies on the Structure of Time: From Physics to Psycho(path)ology
Edited by Buccheri *et al.*, Kluwer Academic / Plenum Publishers, New York, 2000.

signal-processing from these areas remains a focus of theoretical discussion;

b) the problem of *inter-modal* integration, one that isn't necessarily related to cortical mechanisms, since in many species[5] inter-modality is a function of the superior colliculus. In humans this subcortical structure has a relatively small size, with more limited function, the control of eye movements. This fact justifies the association between cortical integration and perceptual binding for humans, but the generalization to non-primates is inadequate;

c) the problem of *psychological* binding, i.e., how different aspects of perception (e.g., in visual perception, form, color and movement), processed by distributed cortical systems, are unified in a single perception. Treisman identified seven different types of perceptual binding in the visual system:
   - binding of properties (shape, color) to objects
   - binding of the parts of an object against a background
   - binding of particular values on a dimension (e.g., ranges of light frequency that are bound into the same color)
   - hierarchical binding, e.g. binding of boundaries to surfaces
   - conditional binding, where "the interpretation of one property (e.g. direction of motion) often depends on another (e.g. depth, occlusion or transparency)"
   - temporal binding of successive states of the same object
   - location binding of objects to their current locations.

The formulation of the "binding" problem in cognitive neuroscience usually conflates physiological and psychological concepts. For example, Gegenfurtner[6] says that "human vision is commonly divided into low level and high level processes. At the lower level, feature analysis occurs to extract information about edges, colors or depths at a particular location in the visual field. At a higher level, the visual system must recognize the objects that are built up from these low feature levels. But there is also an intermediate stage, which presents a major challenge to any theory of vision: how does the visual system know which features belong to which object? ... In order to link features into objects, signals from different parts of the retina, mapped onto different parts of the retinotopic cortical area, must be bound together".

In this work, we discuss the *neuronal* mechanisms of perceptual binding. From psychology we adopt the following common-sense assumptions:

a) perception is unified, i.e. all modalities and aspects of modalities are fused together in an undivided whole;

b) the timing of perceived processes in each modality is generally integrated and consistent with the others, with some well-known exceptions (e.g. the image of lightning appears before the sound).

2. ELECTRICAL OSCILLATORY SYNCHRONY

Discoveries about thalamo-cortical connectivity and columnar specialization, from the fifties to the seventies, led to a current view, that the integration of cortical activity is made by the thalamic-reticular system. In fact, there are many bidirectional (afferent and efferent) connections between practically all cortical regions and the thalamus. The afferent connections project to cortical layer IV, and the efferent connections depart mostly from cortical layers V-VI. Many of the axons of pyramidal cells are directed downwards, suggesting that the preferential direction for the flow of energy/information would be toward the deeper layers and finally the thalamus. The reticular-thalamic system induces rhythms of neuronal firing in the cortex, by the release of excitatory and inhibitory transmitters. Cortico-cortical connections and inter-neurons are also necessary to generate cortical rhythms[7]. However, some of these rhythms are not completely endogenous; they depend on the presence of stimuli and therefore are related to perception.

In the study of evoked potentials, oscillatory synchrony at the gamma (30-50 Hz) frequencies - measured by single and multiple-unit electrodes in experimental animals, and by non-invasive scalp EEG in humans - has been suggested to be a central component of the binding mechanism[8–14]. Engel (ref. 15, p. 220) say that "synchrony may be the 'glue' that binds distributed neuronal activity into unique representations". But how does synchrony work as a 'glue'? The authors say that "in this conceptual framework...synchrony among distributed groups is the relevant code for binding, whereas the intrinsic temporal structure of the responses is, as such, not assumed to represent particular aspects of the visual input" (idem). A possible interpretation of synchrony as "code for binding", proposed by the authors, is that "oscillatory signals may be well suited as carrier signals for a temporal code". In other words, although there is no observed isomorphism between the temporal structure of spikes and properties of stimuli, evoked potentials could "encode" such properties using the synchronous wave as a "carrier" that broadcasts the signals processed by each responsive neuron to the whole assembly.

A second interpretation of the role of oscillatory synchrony is the idea of a "pacemaker" coordinating the activity of large neuronal assemblies. Such a coordination would putatively contribute to produce a state corresponding to a unit of perception[1] developed the hypothesis that perception is "temporally segmented by sequential processing units of 30 to 40 (more conservative: 20-50) ms duration", reflecting states "brought about by neuronal oscillations... Successive system states can be linked together by an integration mechanism. Such temporal integration has an upper limit of a few (approximately 3) seconds". This view implies that all modalities are integrated into 20-50 ms "chunks", which are temporally linked into meaningful sequences with larger duration. But could the minimal unit be the same for all modalities? Comparing the visual and auditory modalities, Poppel et al. say that the *"leading sense"* sets off a relaxation oscillation, defining "a system state for *both* sensory modalities". However, they don't give empirical support to the claim that oscillations triggered by different modalities would have the *same* period.

Another possibility related to the pacemaker interpretation should be brought to analysis. If different modalities entrain different oscillatory periods, there would not exist a common unit of perception. In this view, inter-modal integration would occur in larger temporal intervals, where the synchronizing peaks entrained by the dominant modality would work as *reference pulses* to coordinate all modalities. In this case, the timing of the whole perception would be defined by the timing of the dominant modality. Assuming that a moment of perception has a temporal duration between 100ms and 3s (depending on the difficulty of the perceptual task), for gamma oscillations around 50 Hz (related to modalities with faster percept formation and decay) it would take between 5 and 50 synchronizing pulses to generate an unitary state. The problem with this interpretation is that the temporal dynamics of electrical patterns is being thought of as responsible for too many functions: transduction of signals from stimuli, integrating information within each modality, integrating all modalities and more. Perhaps the complexity of multi-electrode brain measurements reflects such a overload of function.

A third possibility would be that synchronous electric oscillations don't have a critical role for perceptual binding. Their function would be related to sensorimotor functions (one of the possibilities suggested by Roelfsema[13] and one of the results obtained by Miltner[16]), a hypothesis supported by recent advance in the treatment of Parkinson's disease – the installation of a cerebral electric pacemaker that simulates thalamic function is able to eliminate the disease's characteristic motor symptoms. More to the point, Roelfsema et al. suggest that "multiple stimuli may induce responses in pools of neurons that are related to incompatible movements, and which engage in competitive interactions". In order to avoid excessive conflict "neurons that are related to compatible movement components should be linked together with synchronizing connections".

Synchronous oscillations would even contribute to "erase" information in neurons which have already participated in the perceptual process. In this case, perceptual binding would be performed by a different mechanism, that works *before* synchronous oscillations appear. In some cases reported by Pulvermuller[11], gamma oscillations appear 300 ms after the onset of stimuli, while the perception of the stimulus may have occurred between 100-300 ms.

The measurement of synchronous oscillations has surely brought new possibilities to model cognitive functions in the brain, but more experimentation seem to be necessary to explain their function. At the present there is not sufficient evidence to disprove their role for perceptual binding, but it should be stressed that high-frequency oscillatory synchrony, although elicited by reticular-thalamic signaling, is basically a neocortical phenomenon. The binding mechanism should be more general, e.g. it should also be present at least in the superior colliculus. However, other frequencies have also been measured, as the theta (4-7Hz) coherence between prefrontal and posterior association cortex in humans[17]. Since theta rhythms are also found outside the neocortex (e.g. in the hippocampus), combining gamma and theta would increase the strength of the "binding by synchrony" hypothesis.

3. NEUROCHEMICAL OSCILLATORS

Signal processing in the nervous system involves two kinds of patterns: electrical and chemical. Electrical patterns are the result of neuron "firing", producing "spikes" which are directly controlled by the activation and blockage of ion channels. The role of electrical pulses in the cases of signal transmission from peripheral sensors and motor command to muscles is clear; what goes on in-between, i.e., the *cognitive* role of electrical patterns, is another question to be discussed. At the CNS the apparent function of spikes is only to transmit a signal to neurons which are distant from the firing neuron. Therefore, it seems that the understanding of neuronal processes could not occur by studying electrical patterns alone. Following an opposite pathway, experimental studies of perceptual processes have mainly focused on electrical measurements, while neurochemical processes have been largely studied *in vitro*, related to learning and memory functions. Such studies are not, of course, the most appropriate to investigate perceptual and other conscious processes.

The study of neurochemical networks is a new and exciting area of research in neuroscience. One of the lines of research that shed light on molecular mechanisms was the study of the withdrawal reflex in *Aplysia*.

This study suggested that long-term memory would involve gene regulation and dendrite growth by the action of neurotrophic factors[18,19]. Studies of second messengers revealed the existence of three kinds of signal transduction pathways, based on G-proteins, calcium metabolism and protein tyrosine kinases[20]. As the pathways have common elements, computer simulations of their interactions[21] as well as experimental models[22] can be studied. Neurochemical networks present oscillatory rhythms[23] that can be considered the ultimate source of measured electrical patterns; the last ones can be considered a kind of "coarse-grained" effect of neurochemical oscillators, in the sense that the firing of a neuron corresponds to a summation of several molecular processes. Therefore, neurochemical dynamics is far more richer than neuroelectric dynamics, being able to help explaining phenomena intractable in terms of electrical patterns alone.

Studies of intra-neuronal signal-transduction have also revealed the existence of processes at the millisecond range (for $Ca+$ signaling see ref. 24), plausibly related to perceptual processing. Such processes are commonly studied for their role in learning and memory, but despite an apparent dichotomy between perception and learning/memory in scientific methodology, a continuity between them in normal subjects has never been doubted in the literature.

We suggest that perceptual binding could be performed by intra-cellular signal transduction. One difference between perceptual signal transduction and long-term potentiation (LTP) is the *time scale* of the phenomenon; a second one would be the place in the brain where it is assumed to occur (primary sensory cortex instead of the hippocampus); a third lies on the kind of signal that is assumed to open the NMDA channel (see below); a fourth would be the habituation time, assumed to be shorter in perceptual processes; and the final difference would be the substrate putatively phosphorilated by calcium entry.

4. AN ALTERNATIVE HYPOTHESIS

There is a growing consensus between cognitive neuroscientists that the brain works in a *recurrent* neuronal architecture. In the perceptual system, sensory signaling progressively reaches several brain areas, which send feedback signals ("reentrant signaling", in the words of Edelman[25]) to the former ones. In this context, a coordinating mechanism should be sufficiently flexible to the point that a sensory area could identify an afferent signal *and the corresponding reentrant signal* as relative to the same object.

The structure of intra-cellular signal transduction pathways seems to be particularly suitable for this purpose. The activation of AMPA and NMDA channels by thalamic high-frequency excitatory input begins a short-duration cycle (as in the standard hippocampal LTP model reviewed in (ref. 26, p. 35)) that is completed a few milliseconds later with calcium ions entry through NMDA channels. At the primary sensory cortex, one of the factors that may contribute to sustained tetanization is the afferent input's corresponding cortico-cortical reentrant signal.

In our model, perception occurs only after Mg+ is withdrawn from NMDA channels[27] and the entering calcium phosphorilates specific substrates in the cell[28]. Unlocking the NMDA channel would then require the cooperation of afferent and reentrant excitatory signals upon the apical dendrite of a pyramidal cell in primary sensory cortex. Reentrant signaling may come from a variety of associative areas, as the parietal, temporal and pre-frontal areas in primates. Such areas would define the sensory assemblies mobilized to produce a unified perception.

A key point is that such a cooperation should occur within a flexible time interval: a synchronous signal or a late signal would both be ineffective. The disruption of such a delayed coincidence is possibly related to psychiatric phenomena as hallucination. The time interval that defines a perceptual moment is relative to at least four parameters:

a) the interval for the stimulus signal to arrive at the CNS, that is different in different modalities;
b) the interval for the corresponding reentrant signal to reach the sensory areas, that may be different depending on the source of the reentrant signal and the length (number of synapses) of the feedback pathway;
c) the interval necessary for calcium entry and protein phosphorylation; this interval may be different in neurons from different sensory areas, as an adaptation to different temporal patterns coming from the peripheral sensors;
d) the interval necessary to generate a coherent state among the neurons involved in the perceptual process.

Therefore, to define an invariant time unit for perception is not a feasible task. Different time units are possible within a (relatively) broad range, depending on the active sensory modalities and on the difficulty of the stimulus to be processed. The lower limit would be around 100 ms, while the upper limit according to Poppel[1] would be around 3 seconds.

Modality dominance may be based on phylogenetic structural constraints, as the dominance of olfaction in rodents, and/or attentional mechanisms, as the temporary dominance of hearing in humans. Presumably the phylogenetic development of such cellular mechanisms responded to the necessity of avoiding entropy-increasing conflict in a distributed system. The mechanism would be the result of two successive evolutionary strategies[29]: 1) the development of a variety of sensors, providing information for adaptive behavior; 2) integration of the signals processed by the sensors, avoiding dangerous entropy increase in the case of conflicting information. As a result of such an evolution, intra-neuronal metabolic oscillators presumably exist in all structures presumed to perform perceptual binding, as the cat superior colliculus[5].

In summary, we propose that the time "envelope" necessary to complete perceptual processes in primary sensory areas would be defined by an intra-cellular metabolic network in neurons pertaining to the dominant modality. Sensory information from other modalities would be "compressed" or "decompressed" to the size of such an envelope, making a unified perception possible.

REFERENCES

1. Poppel E., Ruhnau E., Schill K., and Steinbuchel N.V., 1990, A Hypothesis Concerning Time in the Brain, *Synergetics of Cognition*, Haken And Stadler (eds.), Heidelberg, Springer-Verlag.
2. Lorig T.S., 1999, On the Similarity of Odor and Language Perception, *Neuroscience and Biobehavioral Reviews*, **23** 391-398.
3. Crick F., 1984, Function of the Thalamic Reticular Complex: The Searchlight Hypothesis, *Proceedings of the National Academy of Sciences*, **81** USA, 4586-4590.
4. Hardcastle V.G., 1994, Psychology's Binding Problem and Possible Neurobiological Solutions, *Journal of Consciousness Studies*, **1** (1) 66-90.
5. Stein B.E. and Meredith M.A., 1993, *The Merging of the Senses*, Cambridge, MIT Press.
6. Gegenfurtner K., 1998, Visual Psychophysics: Synchrony in Motion, *Nature Neuroscience*, **1** (2) 96-99.
7. Llinás R.R., 1992, Oscillations in CNS Neurons: A Possible Role for Cortical Interneurons in the Generation of 40-Hz Oscillations, *Induced Rhythms in the Brain*, Basar and Bullock (eds.), Birkhauser, Boston.
8. Gray C. and Singer W., 1989, Stimulus-Specific Neuronal Oscillations in Orientation Columns of the Cat Visual Cortex, *Proceedings of the National Academy of Sciences*, USA **86** 1698-1702.
9. Llinás R.R. and Ribary U., 1992, Rostrocaudal Scan in Human Brain: A Global Characteristic of the 40-Hz Response During Sensory Input, *Induced Rhythms in the Brain*, Basar and Bullock (eds.), Birkhouser, Boston.
10. Llinás R.R. and Ribary U., 1994, Perception as an Oneiric-like State Modulated by the Senses, *Large-Scale Neuronal Theories of the Brain*, Koch and Davis (eds.), Cambridge, MIT Press.
11. Pulvermuller F., Birbaumer N., Lutzenberger W., and Mohr B., 1997, High-Frequency Brain Activity: Its Possible Role in Attention, Perception and Language Processing, *Progress in Neurobiology* **52** 427-445.
12. Rodriguez E., George N., Lachaux J., Martinerie J., Renault B., and Varela F.J., 1999, Perception's Shadow: Long-Distance Synchronization of Human Brain Activity, *Nature*, **397** 430-433.

13. Roelfsema P.R., Engel A.K., Konig P., and Singer W., 1996, The Role of Neuronal Synchronization in Response Selection: A Biologically Plausible Theory of Structured Representations in the Visual Cortex, *Journal of Cognitive Neuroscience*, **8** (6) 603-625.
14. Singer W., 1993, Synchronization of Cortical Activity and Its Putative Role in Information Processing and Learning, *Annual Review of Physiology*, **55** 349-374.
15. Engel A.K., Konig P., Schillen T.B., and Singer W., 1992, Temporal Coding in the Visual Cortex: New Vistas on Integration in the Nervous System, *Trends in Neuroscience*, **15** (6) 218-226.
16. Miltner W.H.R., Braun C., Arnold M., Witte H., and Taub E., 1999, Coherence of Gamma-Band EEG Activity as a Basis for Associative Learning, *Nature*, **397** 434-436.
17. Sarnthein J., Petsche H., Rappelsberger P., Shaw G.L., and Stein A. von, 1998, Synchronization Between Prefrontal and Posterior Association Cortex During Human Working Memory, *Proceedings of the National Academy of Sciences*, **95** 7092-7096.
18. Bailey C.H. and Kandel E.R., 1995, Molecular and Structural Mechanisms Underlying Long-Term Memory, *The Cognitive Neurosciences*, Gazzaniga (ed.), Cambridge, MIT Press.
19. Black I.B., 1995, Trophic Interactions and Brain Plasticity, *The Cognitive Neurosciences*, Gazzaniga (ed.), Cambridge, MIT Press.
20. Rocha A.F., 1997, The Brain as a Symbol-Processing Machine, *Progress in Neurobiology*, **53** 121-198.
21. Bhalla U.S. and Iyengar R., 1999, Emergent Properties of Networks of Biological Signaling Pathways, *Science* **283** 381-387.
22. Park D. and Dunlap K., 1998, Dynamic Regulation of Calcium Influx by G-Proteins, Action Potential Waveform and Neuronal Firing Frequency, *The Journal of Neuroscience*, **18** (17) 6757-6766.
23. Goldbeter A., 1992, Comparison of Electrical Oscillations in Neurons with Induced and Spontaneous Cellular Rhythms due to Biochemical Regulation, *Induced Rhythms in the Brain*, Basar And Bullock (eds.), Birkhauser, Boston.
24. Alkon D.L., Nelson T.J., Zhao W., and Cavallaro S., 1998, Time Domains of Neuronal Ca2+ Signaling and Associative Memory: Steps Through a Calexcitin, Ryanodine Receptor, K+ Channel Cascade, *Trends in Neuroscience*, **21** (12) 529-537.
25. Edelman G.M., 1989, *The Remembered Present: A Biological Theory of Consciousness*, Basic Books, New York.
26. Bliss T.V.P. and Collingridge G.L., 1993, A Synaptic Model of Memory: Long-Term Potentiation in the Hippocampus, *Nature*, **361** 31-39.
27. Flohr H., 1995, Sensations and Brain Processes, *Behavioral Brain Research*, **71** 157-161.
28. Rocha A.F. and Pereira Jr. A., (forthcoming) NMDA Signal Transduction Pathways and Consciousness. Submitted for publication.
29. Pereira Jr. A. and Rocha A., 1999, Coerência, Entropia e Consciência. To appear in Gonzales, M.E.Q. and D'Ottaviano, I.M. (eds.) *Auto-Organização: Estudos Interdisciplinares II*, Campinas, Centro de Lógica e Epistemologia.
30. Tononi G., Edelman G.M., and Sporns O., 1998, Complexity and Coherency: Integrating Information in the Brain, *Trends in Cognitive Science*, **12** (2) 474-484.

Do it with rhythm

How internal clocks can simplify life for artificial and biological organisms

[1]DAVIDE MAROCCO, [2]ORAZIO MIGLINO, AND [2]RICHARD WALKER
[1]*Consiglio Nazionale delle Ricerche, Roma*
[2]*Università di Napoli II*

1. INTRODUCTION

The theme of this workshop is time. In this paper we will examine the role of time, and the ability to measure time, in certain very simple forms of animal cognition and behavior.

It is well known that virtually all animals possess "internal clocks" operating on a variety of scales - from milliseconds to months or even years[1]. These clocks - some of whose molecular and genetic mechanisms have been identified by recent research - regulate a broad variety of behaviors and body rhythms, from sleep and wakefulness to reproduction and migration[2]. Time and rhythm are, in short, a basic component of animal biology.

Internal clocks modulate the dynamics of neurotransmitter availability in the brain[3,4]. It is reasonable, therefore, to assume that cognition can make use of clock data. Here we use techniques from "Artificial Life"[5] to show that "artificial organisms" with "internal clocks" are able to generate apparently complex behaviors using extremely simple neural mechanisms which did not require the complex "internal representations" posited by alternative models of the same behavior. This suggests that ability to measure the passage of time is strongly advantageous to animals, greatly simplifying the cognitive and behavioral tasks they are asked to perform.

Studies on the Structure of Time: From Physics to Psycho(path)ology
Edited by Buccheri *et al.*, Kluwer Academic / Plenum Publishers, New York, 2000.

2. TWO SIMPLE BEHAVIORS

In our experiments we "evolve" "artificial organisms" exhibiting two well-known animal behaviors: the ability to "wander efficiently"[6,7] and to "detour"[8,9].

The first behavior we examine is *efficient wandering*. When a mouse is released into an unfamiliar environment it "wanders" in what seems to be a random fashion. "Optimal foraging theory" predicts, however, that wandering will be "efficient": on release the mouse will rapidly visit a large proportion of the available territory[10]. The mouses's behavior is, in other words, less random than it seems. A number of experimental studies have supported this hypothesis[11].

The second behavior we examine is the classical "detour problem". When an animal seeks to reach food or a mate it will often find an obstacle in the way. In these circumstances the only way it can reach its goal is to take an indirect route, involving the loss of visual contact with the target. This is "a detour". Detours have been demonstrated in many animals including chimpanzees[12] and rats[13,14]. In the experiment reported here we replicate the detour behavior of two day old chicks, as observed by Regolin et al.[15]

3. THE EXPERIMENTS

It is well-known that manual modeling of complex behaviors can be extremely difficult and often leads to proposals which are "un-biological" in flavor. In our experiments we avoided these difficulty by using a so-called Genetic Algorithm (GA)[16,17] to "evolve" populations of "artificial organisms" (see methods section) controlled by Artificial Neural Networks (ANNs) and displaying the ability to perform a set task: in our first set of experiments "efficient wandering", in the second set: a detour.

Both sets of experiments involved the evolution of a population of 100 such "organisms". At the beginning of the experiment we generated random genomes for each organism. At the end of each generation we measured organisms' ability to perform the set task. Each was assessed in a number of different settings. In the case of wandering we used a "fitness function" which rewarded completeness and speed of exploration (see methods section). In the case of detour the fitness formula rewarded robots which successfully searched for the target and moved towards it. At the end of the testing process the 20% of fittest organisms were "selected" for reproduction. Each of the selected organisms was made to produce five clones. During this process we introduced random "mutations": bits in the genome were flipped from 0 to 1 or vice versa with a probability of 0.04 per bit/generation (wandering) and 0.02/bit/generation (detour). The cycle of test, reproduction and mutation was iterated respectively for 600 generations (wandering) and 350 generations (detour).

Finally we observed the behavior of organisms in a "novel environment" (see methods section) which they had not met during the evolutionary process. This test made it possible to measure the robustness of evolved network configurations with respect to minor changes in task definition.

4. RESULTS

In both sets of experiments we successfully evolved organisms capable of achieving high levels of performance in novel environments as well as in the specific settings where they had evolved.

4.1 Wandering

In our work on wandering behavior we compared the performance of organisms with and without "internal clocks". The results showed that so long as organisms were tested in the same environment where they had evolved they achieved similar levels of fitness - whether or not they possessed an internal clock. In novel environments, on the other hand, organisms with internal clocks outperformed organisms without clocks (see Figure 1).

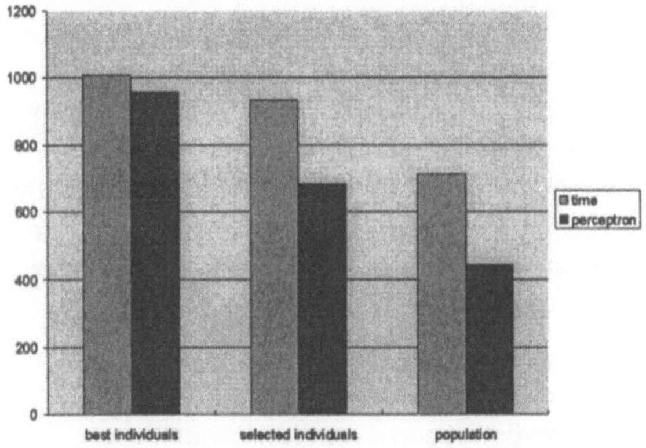

Figure 1. Wandering behavior in a rectangular "open field box". Trajectory of a robot with "internal clocks".

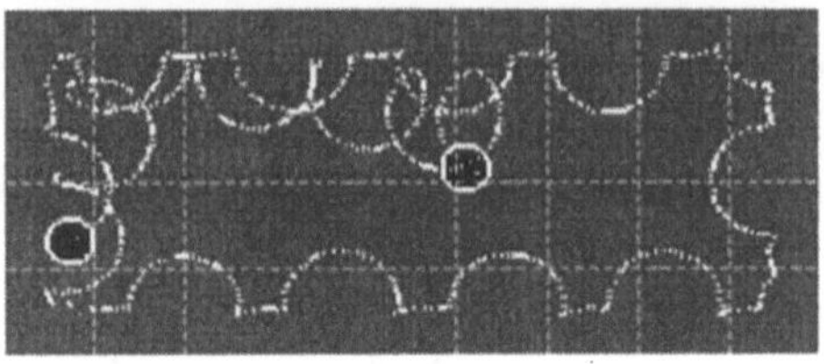

Figure2. Wandering in a rectangular "openfield box": trajectories followed by robots with no "internal clocks".

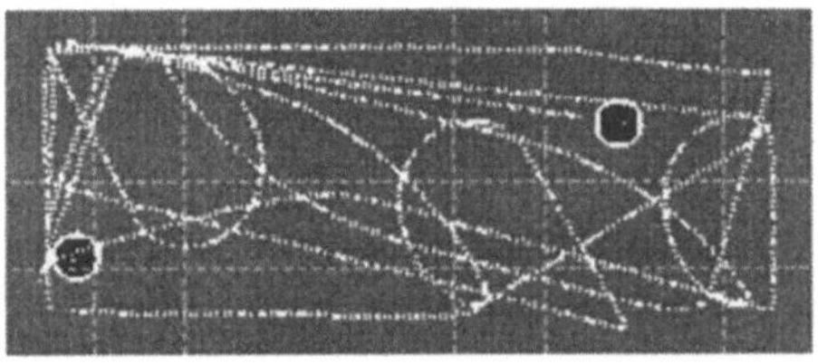

Figure3. Efficiency of wandering in robots, with and without "internal clocks".

An examination of the trajectories followed by these two classes of organism helps to explain these results. Figure 2 shows a typical trajectory for an organism with no internal clock. This strategy is highly adapted to the specific environment where the organism has evolved. The diameter of the semi-circles and the length of the line segments in the trajectory are perfectly adapted to the cell-size used for evaluation purposes. The easy to describe, environment-dependent strategies evolved by organisms with no internal clock may be compared with the more complex behavior produced by organisms which are able to take account of time input (see Figure 3). In experiments with these organisms we observe, as in the previous case, that the organism turns on meeting an obstacle. It appears however that it also turns (at an oblique angle) on receiving input from its internal clock. In many cases the robot will then traverse the field box following a curved trajectory which maximizes the number of cells visited en route. The ability to periodically change direction is independent of the external environment where the robot has evolved.

Comparing organisms with and without "internal clocks" it may be observed that in the absence of any other source of input the former have necessarily adapted to the specific environments where they evolved. The latter, on the other hand, have access to a source of input (the "internal clocks") whose characteristics are independent of the external environment. This explains the superior performance of "clocked" organisms in unfamiliar environments.

4.2 Detour

In the case of detour behavior systems without internal clocks completely failed to generate effective detours; with internal clocks they behaved in ways which were statistically indistinguishable from the behavior of the two day old chicks used by Regolin et al.[15] in their experiments[6,8]. Here again an examination of trajectories (see Figure 4) casts light on the way in which a "sense of time" can assist behavior.

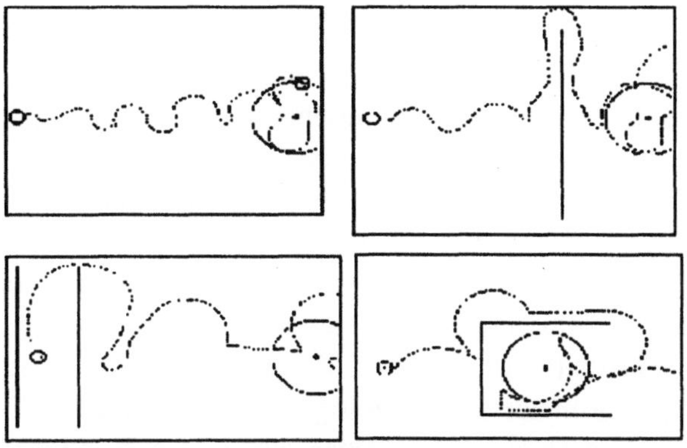

Figure 4. Detour behavior: trajectories followed by a typical robot in the four training environments. Legend: Small circles: terminal points for trajectories (if within field box). Large circle: target area rewarded by fitness function.

In different simulations robots evolved different strategies. We have not as yet performed a detailed analysis of these strategies or of the underlying computational mechanisms. In general however they appear to be based on simple rules, for example (see Figure 4):

1. Main: Move forward turning first clockwise (slowly) and then anti-clockwise (more rapidly).
2. Taxis: On visual contact with target turn sharply towards the target. Return to Main
3. Wall following: If left proximity sensors active turn right. If right sensors active turn left. Move forwards until obstacle out of view. Return to Main

(1) - an essential component in the overall strategy - requires a sense of "time". When the robot is a long way from any obstacle it receives no external stimuli. It is only the input from its "internal clocks" which allows it to modify its behavior, actively exploring the environment for new stimuli. The combination of exploration and taxis, which these sensors make possible, represents an efficient strategy for moving towards the target even when it is often outside the camera's field of vision.

5. DISCUSSION AND CONCLUSIONS

Classical models of animal behavior make no explicit reference to time. For the behaviorist animal behavior is a simple function of the stimuli the animal receives from the outside world[18]. Schematically:

$$B = f(S) \tag{1}$$

where B is the behavior and S a set of external stimuli.

This is the behavior we observe in "artificial organisms" with no internal clock. If the environment is stable "artificial evolution" can achieve impressive results, as in our experiments with wandering in a familiar environment. Optimal behavior in one environment may however be severely suboptimal in a another. Our experiments show that an organism may evolve a highly effective strategy for wandering in an open field box of one particular shape; a small change in the measurements of the box is enough however to completely invalidate the strategy. For anyone who has observed the elasticity of real-world animal behavior such a rapid degradation of performance seems biologically unrealistic.

Modern cognitive science recognizes that in reality animal behavior is more than a simple response to external stimuli. Many behaviors are best explained by positing the existence of "internal representations". Schematizing again:

$$\begin{aligned} B_t &= f(S_t, R_t) \\ R_t &= g(S_{t-1}, R_{t-1}) \end{aligned} \tag{2}$$

where B_t, S_t and R_t stand for Behavior, Stimuli and Internal representations at time t. These are the equations for a "finite state automaton". Many cognitive scientists argue that navigational skills (such as those needed for efficient wandering or detours) require the presence of "topographical maps"[11,13,14], perhaps implemented via "place cells" within the animal's brain[19]. In our own experiments, on the other hand, "topographical maps" or "place cells" are excluded by design; the network architecture has no "hidden neurons"; there is no location where place information could be represented; the only inputs received by the output neurons are those from the sensors and the internal clocks. In short:

$$B = f(S, T) \tag{3}$$

where T is the output from an "internal clock"- a counter which counts up to a certain number and cycles back to zero.

We prove, in other words, that very simple "artificial organisms" with internal clocks can perform efficient wandering and simple detours without resort to internal representations. This does not demonstrate that biological organisms do not possess or use such representations; this is a highly controversial issue[11,13,14,19,20,21]. What it does show is that the use of time data can significantly simplify a number of cognitive tasks. There are many circumstances in which it is useful to change behavior once a certain time has passed - even in the absence of external stimuli. Given that all animals possess internal clocks it would be surprising if evolution had failed to exploit them for cognitive purposes. It seems very likely that "doing it with rhythm" makes life simpler for biological as well as artificial organisms.

6. METHODS

6.1 Artificial organisms

The artificial organisms used in our experiments were software simulations[22] of the well-known Khepera robot[23]. Input to the robot came from 8 infrared proximity sensors, 4 sensors linked to a linear video-camera and 3 "time sensors". Proximity sensors have a sensory field of 20° and are sensitive to obstacles within 3 cm of the sensor. Output (between 0 and 1) is a continuous, decreasing function of distance to the obstacle. The video-camera has a field of vision of 36°. Each sensors produces an output of 1 if the center of the target lies within its own 9° field of vision. The output values of the three time sensors were initially set to zero, increasing respectively by 0.01, 0.02 and 0.03 on each cycle of computation. When a sensor value reached 1 it was reset to zero. The motor apparatus consisted of a left and a right wheel driven by stepping motors which can move both forwards and backwards. The motor apparatus was controlled by an Artificial Neural Network (ANN) with input neurons representing the state of the sensors and output neurons controlling the stepping motors. A number of different architectures were tested. The architecture finally chosen was based on a simple Perceptron[24] in which all sensors have a direct connection to the two output units. Evolution involved "mutations" in connection strengths. The genome of the organism consisted of a sequence of binary coded numbers (8 bits per number) representing the strengths of individual connections.

6.2 Simulation environments

6.2.1 Wandering

Robots were tested four times in a rectangular 90 cm by 40 cm box and four times in a square 60 cm by 60 cm box. On each test the robot started with a randomly chosen position and orientation. For measurement purposes the terrain was divided into square 10 cm by 10 cm cells. Each test consisted of 1,000 cycles of computation.

6.2.2 Detour

Each environment consisted of an open field with no external fence. In the first environment there was no obstacle between the robot and the target. The second, third and fourth environments selected for actual detour behavior. In the second environment the target was placed behind a linear obstacle 80 cm long. In the third environment Khepera was placed inside an 10*80 cm corridor. The fourth environment used a 40*70 cm U-shaped obstacle (see Figure 4). Obstacles were 3 cm high and of negligible thickness; the target was 12 cm high. It follows that in the "evaluation sessions" the robot was always able to "see" the target even when the path to the target was obstructed by an obstacle. The evaluation test was repeated five times for each environment. At the beginning of each cycle the robot was placed in a randomly chosen position 90 cm from the target. The heading was chosen randomly from a uniform distribution.

6.3 Fitness functions

6.3.1 Wandering

$$fit = ncells + (nCycles - cyclesToCompletion)$$

Where:

fit is the fitness attributed to the robot

nCells is the number of different cells touched by the robot during exploration

nCycles is the duration of the exploration (in cycles)

and

cyclesToCompletion is the number of cycles traversed before the robot has touched every cell in the box. (If the robot never touches all the cells *cyclesToCompletion* is assigned the same value as *nCycles*.)

6.3.2 Detour

On each cycle of computation, for each organism the system computes the following function

```
IF ("some infrared sensor" > 0 && old_position >< new_position)
  fitness ++
IF ("some infrared sensor" > 0 && old_position = new_position)
  fitness --
IF (distanceToTarget < 15 cm.)
  fitness += 10
```

The first two components in the fitness formula are designed to encourage obstacle avoidance; the last component rewards the robot when it approached the target.

6.4 Novel environments

6.4.1 Wandering

The novel environment consisted of a 120 cm by 90 cm box which the organisms had not encountered during the evolutionary process. As in the tests of fitness during the evolutionary process each robot was tested from 4 different starting positions. Given the larger size of the new environment the number of computation cycles was increased to 3,000.

6.4.2 Detour

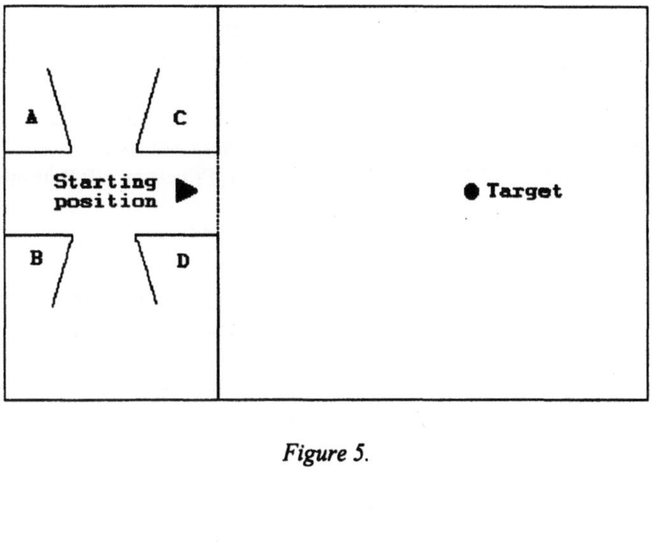

Figure 5.

The organisms were tested in a computer replica of an experimental setting used by Regolin et al. to demonstrate detour behavior in two day old chicks. (see Figure 5)

The organisms are placed in a "cage" divided in two by a barrier. The part of the cage on the other side of the barrier contains a corridor. On the end of the corridor facing the target there is a window through which the organisms can "see" the target. On each side of the corridor there are apertures leading into two compartments, one facing the target, and one facing in the opposite direction. The two compartments facing the target are labeled C and D.; the two compartments facing in the other direction are labeled A and B. At the beginning of each test the organism is placed in the corridor close to the barrier and allowed to wander freely. The system records the first compartment the organism enters and the time it takes to reach it. The test is halted after 600 cycles of computation.

REFERENCES

1. Farner D.S. 1985, Annual Rhytms, *Annual Review of Physiology*, **47** 65-82.
2. Aschoff J., 1981, *Biological Rhythms*, Plenum press, New York .
3. Moore R.Y. and Spek J.C., 1993, GABA is the principal neurotransmitter of the circadian system, *Neuroscience letters*, **150** 112-116.
4. Wagner S., Castel M., Gainer H., and Yarom Y., 1997, GABA in the mammalian suprachiasmatic nucleus and its role in diurnal rhythmicity, *Nature* **387** 598-603.
5. Langton C., 1997, *Artificial Life, an Overview (Complex Adaptive Systems)*, Bradford Books.
6. Walker R. and Miglino O., 1999, Simulating exploratory behavior in evolving artificial neural networks, *Proceedings of the Genetic and Evolutionary Computation Conference.*
7. Miglino O., Nafasi K. and Taylor C., 1995, Selection for wandering behavior in a small robot, *Artificial Life.*
8. Walker R. and Miglino O., 1999, Replicating experiments in "Detour behavior" with artificially evolved robots: an A-Life approach to comparative psychology, *Advances in Artificial Life*, Lecture Notes in Artificial Intelligence 1674, Springer.
9. Miglino O., Denaro D., Tascini E., and Parisi D., 1998, Detour behavior in evolving robots, *Proceedings of First European Workshop on Evolutionary Robotics.*
10. Stephens D.W. and Krebs J.R., 1986, *Foraging theory*, Princeton, NJ: Princeton University Press.
11. Gallistel C.R., 1990, *The Organization of Learning*, MIT Press.
12. Koelher W., 1925, *The Mentality of Apes*, Harcout Brace, New York.
13. Tolman E.C., 1948, Cognitive maps in rats and men, *Psychological Review*, **36** 13-24.
14. Tolman E.C. and Honzik C.H., 1930, Insight in rats, *University of California Publications in Psychology*, vol 4, 77-8, 215-232.
15. Regolin L., Vallortigara G., and Zanforlin M., 1994, Object and Spatial Representations in Detour Problems by Chicks, *Animal Behavior*, **48** 1-5.
16. Holland J.H., 1975, *Adaptation in Natural and Artificial Systems*, 1st ed., Cambridge, MA: the MIT press.
17. Mitchell M., 1996, *Introduction to Genetic Algorithms*, MIT Press.

18. Watson J.B., 1925, *Behaviorism*, Chicago University Press.
19. O'Keefe J. and Nadel L., 1978, *The hippocampus as a cognitive map*, Oxford University Press.
20. Bennett A.T.D., 1996, Do animals have cognitive maps?, *Journal of Experimental Biology*, **199** (1) 219-224.
21. Burgess N., Donnett J.G., O'Keefe J., 1997, Robotic and Neuronal Simulation of Hippo-campal navigation, *AISB Workshop on "Spatial reasoning in Mobile Robots and Animals*, Technical Report n. UMCS-75-4-1, University of Manchester.
22. Miglino O., Lund H.H., and Nolfi S., 1996, Evolving Mobile Robots in Simulated and Real Environments, *Artificial Life*, **2** (4) 417-434.
23. Mondada F., Franzi E. and Ienne P., 1993, Mobile Robot miniaturisation. A tool for investigation in control algorithms. *Proceedings of the Third International Symposium on Experimental Robotics*, Kyoto, Japan.
24. Minsky M. and Papert S., 1988, *Perceptrons*, MIT Press.
25. Nolfi S., Floreano D., Miglino O., and Mondada F., 1994, How to evolve autonomous robots: different approaches in evolutionary robotics, *Artificial Life IV - Proceedings of the Fourth International Workshop on the Synthesis and Simulation of Living Systems*, Brooks R.A. and Maes P. (eds.), Cambridge MA, MIT Press, 190-197.

2.

QUANTUM PHYSICS AND ARROW OF TIME

Pregeometry and the Trans-Temporal Object

MARK STUCKEY
Dept. of Physics, Elizabethtown College, Elizabethtown, PA, USA 17022

1. INTRODUCTION

Fundamental theoretical physics is in a seriously stalled state[1]. An underlying cause for the quagmire of theoretical physics is a tacit adherence to the dynamical perspective (dynamism). Dynamism holds that fundamental reality includes objects with trans-temporal identity and their causal interactions mediated by the four forces carried by their respective elementary particles as modeled in a local spacetime setting. Dynamism also assumes presentism - the spacetime manifold is foliated via an equivalence relation (simultaneity) and time is a one-dimensional linearly ordered quotient set induced by simultaneity.

Arguments can be presented against each of the defining features of dynamism[1]. The list of such arguments is well-established in the philosophy of science.

- Relativistic quantum field theory and quantum non-locality (QNL) against trans-temporal objects.
- QNL and special relativity (SR) against causality.
- Multiplicity iff discernibility[2] against the ontology of force.
- The relativity of simultaneity against presentism.

Let me summarize two of these arguments here. In deductive form, we have the following argument against presentism.

P1. According to presentism, only the present is real. The past and future are not real.

P2. According to SR there is no way to identify unique planes of simultaneity.

P3. According to SR, the present is a plane of simultaneity.

Studies on the Structure of Time: From Physics to Psycho(path)ology
Edited by Buccheri *et al.*, Kluwer Academic / Plenum Publishers, New York, 2000.

C. SR does not allow for the unique spatio-temporal foliation required for presentism.

Philosophers have posed counter arguments and rebuttals *ad nauseam*, but this argument is widely regarded as convincing[3]. Presentism is a naive perspective in light of SR.

The other argument I briefly summarize is QNL against trans-temporal objects. By QNL I mean correlated, space-like separated experimental outcomes which violate Bell's inequality. QNL was introduced as early as 1935 in a paper by Einstein, Podolsky, and Rosen[4] and is sometimes referred to as the EPR paradox. In 1964, Bell[5] explained how the EPR conundrum is more than a metaphysical issue when he showed there are realizable consequences. And, in the early 1980's Aspect *et al* published experimental results[6] confirming the violation of Bell's inequality and validating the predictions of quantum mechanics (QM).

The violation of Bell's inequality implies a non-local connection or 'collusion' between particles with correlated properties. This jeopardizes our picture of interacting trans-temporal objects, since implicit in the construct of such objects is the local association of events via the tangent space to the spacetime manifold. As an analogy of the local mapping via tangent vectors on the spacetime manifold, consider the construct of "a man's image" on a film strip. We map right index finger to right index finger and left thumb to left thumb, frame by adjacent frame. But QNL with its correlated space-like separated events, suggests a truly fundamental description would allow for a point on the right index finger of frame 10 to be correlated with a point on the left thumb of frame 8.

That is, QNL suggests we distinguish between the neighborhoods of events in microscopic systems and the neighborhoods of events in macroscopic systems. [Note: In this context, the term "event" refers to an element of the spacetime manifold.] When considering many events, we expect the notion of locality to be commensurate with the notion of an open ball *a la* calculus on manifolds. In this context, we may construct trans-temporal objects and employ the dynamical perspective. But if considering few events, the notion of locality might be defined so that when placed in the macroscopic context, microscopic neighborhoods may exhibit non-locality. Therefore, the modeling of spacetime might have to be reconsidered at a fundamental level. Demaret et al write[7],

In this Section we analyze fundamental concepts of quantum mechanics. We show that they lead to some problems which force us to modify the usual notion of space-time. ... The second problem is related to the famous E.P.R. paradox which introduces the idea of non-locality or more precisely of non-separability with respect to space. In fact, in quantum mechanics space

cannot be viewed as a set of isolated points. These problems lead to a deep modification of our representation of "quantum" space-time.

Perhaps the most ambitious programs in the attempt to model quantum spacetime are those of pregeometry[8]. According to pregeometry, "the features of the conventional spacetime, such as its continuity, dimensionality, and even causality and topology, should not be present from the beginning, but should emerge naturally in the transition process from pregeometry to the usual spacetime dynamics of our conventional physical theories." Pregeometric programs are well-suited for addressing QNL, since they are *ipso facto* unadulterated by conventional notions of locality.

As a caveat, some pregeometric programs are not strictly reductive. By strictly reductive I mean that concepts/phenomena are described deductively and hierarchically from a set of undefined objects and a collection of axioms/postulates. There are pregeometric programs for example in which fundamental entities are dynamic, e.g., momentum or energy. Such programs necessarily harbor circularity. Thus, one cannot employ dynamical concepts at the pre-Riemannian (non-metric) levels of modeling in a strictly reductive pregeometric program. Reductive pregeometry is preferred, as circularity in descriptive systems is typically eschewed.

To illustrate the feasibility of reductive pregeometry I employ uniformities on the sets underlying discrete topological groups. The fundamental elements of the formalism herein are elements of a denumerable set Y. Elements of Y x Y form events of the spacetime manifold per their membership in the entourages of a uniform space. In this context, macroscopic and microscopic spacetime neighborhoods can be clearly defined and delineated. Accordingly, locality is understood conventionally via macroscopic spacetime neighborhoods and non-locality obtains when microscopic neighborhoods are viewed as subsets of macroscopic neighborhoods. This toy model provides a mathematical articulation of how the fundamental elements of pregeometry displace the trans-temporal objects of dynamism as fundamental elements in the modeling of reality.

2. THE FORMALISM

In constructing a definition of spacetime neighborhoods, we do not want to employ mathematics which is steeped in a notion of locality *a priori*. Thus, we avoid differential calculus to include the differential manifold[9], as well as notions such as the field structure over the (otherwise structureless) set of reals. We wish to start with an unordered set and develop explicitly the 'placement' of its elements in a mathematically natural fashion[10]. We also seek a formalism which is independent of the domain of discourse for the underlying set *a la* general covariance.

A uniform space can be produced via any topological group by constructing a set of entourages from neighborhoods of the identity. And, a topological group can be constructed over any group by choosing the discrete topology over the underlying set[11]. So, we are guaranteed that a uniform space may be constructed over a nondescript set by introducing a group structure and the discrete topology over it. Concerning denumerable sets, Z_N is a group structure for all N and is uniquely so when N is prime. Thus, we are motivated to investigate the 'skeletal' entourage structure over the topological group Z_N.

3. ENTOURAGES VIA THE TOPOLOGICAL GROUP Z_N

Let us denote by Y the underlying set and by Z_N the set with algebraic structure. Z_N is cyclic and I denote by 'a' its generator, so the identity is a^N. Entourages are constructed from pairs $(b,c) \in Y \times Y$ according to the axioms of a uniform space. The collection of all pairs $(b,c) \ni b = c$ is called the *diagonal* and is denoted D. The diagonal is axiomatically and uniquely a subset of every entourage. Entourage multiplication is also defined. For A and B entourages, AB is the entourage containing all pairs $(b,d) \ni (b,c) \in A$ and $(c,d) \in B$. Since $D \subset A$ and $D \subset B$, $A \subset AB$ and $B \subset AB$.

To construct what I refer to as the "skeletal" set of entourages, we consider the smallest neighborhoods of the identity. Denote these elements of the topology by $\{a^N, a^i\}$ with $i = 0, 1, 2, ..., N-1$. An entourage A_i generated from $\{a^N, a^i\}$ is the collection of pairs $(b, c) \ni bc^{-1} \in \{a^N, a^i\}$. Therefore,

$$A_0 = D,$$
$$A_1 = \{D, (a^N, a^{N-1}), (a^1, a^N), (a^2, a^1), (a^3, a^2), ..., (a^{N-1}, a^{N-2})\},$$
$$A_2 = \{D, (a^N, a^{N-2}), (a^1, a^{N-1}), (a^2, a^N), (a^3, a^1), ..., (a^{N-1}, a^{N-3})\}, \text{ etc.}$$

We see that $\forall$ b, c $\in$ Y, the N(N − 1) pairs $(b,c) \ni b \neq c$ are partitioned equally among the A_i. [This is true of the skeletal entourages over any discrete topological group, since any member of the group can be written as an inverse and the group multiplication table contains N unique pairs for any given element of the group.]

4. SPACETIME NEIGHBORHOODS

We seek a natural mathematical differentiation between macroscopic and microscopic spacetime neighborhoods. Let the pairs (b,c) constitute spacetime events *a la* elements of the spacetime manifold of general relativity. Since the A_i partition the pairs (b,c) $\ni$ b $\neq$ c, microscopic spacetime neighborhoods are identified with the A_i. We expect the whole of spacetime to involve collections of microscopic spacetime neighborhoods, so we construct it by entourage multiplication of the A_i. Once we've spacetime *in toto*, we may define macroscopic spacetime neighborhoods via conventional locality. To define the whole of spacetime, I first note a fact about multiplication of the skeletal entourages over a general, discrete topological group.

Let A_α be the entourage generated from {e,α} and A_β be the entourage generated from {e,β}. Consider the N pairs {(b,c), (c,d) | (b,c) $\in A_\alpha$ and (c,d) $\in A_\beta$ with b $\neq$ c and c$\neq$d}. In addition to D, these account exhaustively for the elements of A_α and A_β. For any such pair, (b,d) $\in A_\alpha A_\beta$ by definition of entourage multiplication and (b,d) $\in A_{\alpha\beta}$, since $\alpha\beta = (bc^{-1})(cd^{-1}) = bd^{-1}$. The N pairs (b,d) with D account exhaustively for the elements of $A_{\alpha\beta}$. And, excepting the impact of D on $A_\alpha A_\beta$, the N pairs (b,d) account exhaustively for the elements of $A_\alpha A_\beta$. The impact of D on $A_\alpha A_\beta$ is to render $A_\alpha \subset A_\alpha A_\beta$ and $A_\beta \subset A_\alpha A_\beta$. Therefore, $A_\alpha A_\beta = A_\alpha \cup A_\beta \cup A_{\alpha\beta}$.

So we see that when the underlying group structure is cyclic, multiplication of skeletal entourages builds entourage structure paralleling generation of the group. For the topological group Z_N we have

$$A_1^2 = A_1 \cup A_2\,,$$
$$A_1^3 = A_1 \cup A_2 \cup A_3\,,$$

... and

$$A_1^{N-1} = A_1 \cup A_2 \cup \ldots \cup A_{N-2} \cup A_{N-1}$$

which constitutes all pairs (b,c) $\in$ Y x Y. Thus, it seems natural to define spacetime *in toto* via A_1^{N-1}. A 2-dimensional spacetime structure suggests itself immediately in the form of a simple array,

$$
\begin{array}{ccccc}
(a^1, a^N) & (a^1, a^1) & (a^1, a^2) & \cdots & (a^1, a^{N1}) \\
(a^2, a^N) & (a^2, a^1) & (a^2, a^2) & \cdots & (a^2, a^{N1}) \\
\cdot & \cdot & \cdot & \cdot & \cdot \\
\cdot & \cdot & \cdot & \cdot & \cdot \\
\cdot & \cdot & \cdot & \cdot & \cdot \\
(a^N, a^N) & (a^N, a^1) & (a^N, a^2) & \cdots & (a^N, a^{N1})
\end{array}
$$

In this context, the conventional macroscopic spacetime neighborhood of (a^2, a^1) would be $\{(a^1, a^N), (a^1, a^1), (a^1, a^2), (a^2, a^N), (a^2, a^2), (a^3, a^N), (a^3, a^1), (a^3, a^2)\}$, i.e.,

$$
\begin{array}{ccc}
(a^1, a^N) & (a^1, a^1) & (a^1, a^2) \\
(a^2, a^N) & (a^2, a^1) & (a^2, a^2) \\
(a^3, a^N) & (a^3, a^1) & (a^3, a^2)
\end{array}
$$

However, its microscopic neighborhood is A_1 which when placed in the context of the whole spacetime is highly non-local (omitting D for brevity)

$$
\begin{array}{ccccc}
(a^1, a^N) & & & & \\
 & (a^2, a^1) & & & \\
 & & (a^3, a^2) & & \\
 & & & - & \\
 & & & & (a^N, a^{N1})
\end{array}
$$

And for the microscopic neighborhood of $(a^i,a^j) \in A_2$ we have (again, omitting D)

(a^1, a^{N1})

(a^2, a^N)

(a^3, a^1)

–

(a^N, a^{N2})

5. CONCLUSION

While this is but a toy model, it depicts in a mathematically rigorous fashion the manner in which pregeometry might overcome problems associated with the dynamical perspective. And, the 2-D spacetime is not irrelevant *per se*. It might, for example, lend itself to modeling the time-dependent, horizontal distribution of photon detection events in the twin slit experiment[12].

Of interest here is the manner in which the trans-temporal object is displaced from the set of fundamental elements used to model reality. Rather than describing reality in terms of interacting atoms, molecules, etc., we seek mathematical relationships between objects with no counterparts in reality. Only when large numbers of such 'Platonic' objects are considered, does one expect to obtain a basis for the construct of trans-temporal objects. In this sense, we are "glimpsing into a level of underlying objective reality that is deeper than perceived reality[13]".

REFERENCES

1. Silberstein M. and Stuckey W.M., 1999, "Static for Dynamism: Reductive Pregeometry and Unification", submitted to *J. Phil. Sc.*
2. Stuckey W.M., "Leibniz's Principle, Dynamism and Non-locality" to be published in *Physics Essays.*
3. Davies P., 1983, *God & the New Physics*, Simon & Schuster, New York, 128.
4. Einstein A., Podolsky B., and Rosen N., 1935, *Phys. Rev.* 47, 777.
5. Bell J.S., 1964, *Phys. 1*, 195.
6. Aspect A., Grangier P., and Roger G., 1981, *Phys. Rev. Lett.* 47, 460.
7. Demaret J., Heller M., and Lambert D., 1997, *Foundations of Science* 2, 137.
8. Gibbs P., 1995, "The Small Scale Structure of Space-Time: A Bibliographical Review" hep-th/9506171.

9. Stuckey W.M., 1998, *Causality and Locality in Modern Physics*, Hunter, Jeffers, and Vigier (eds.), Kluwer, Dordrecht, 235.
10. Stuckey W.M., 1997, *Modern Mathematical Models of Time and their Applications to Physics and Cosmology*, Tifft and Cocke (eds.), Kluwer, Dordrecht, 371.
11. Geroch R., 1985, *Mathematical Physics*, University of Chicago Press, Chicago, 234.
12. Stuckey W.M. and Silberstein M., 1999, "A pregeometric approach to modeling quantum non-locality" submitted to *J. Math. Phys.*
13. Rosen J., 1994, *Physics Essays* 7, **3** 1.

Quantum correlations, large scale structure of the universe and temporal non-locality

[1]SISIR ROY AND [1,2]MENAS KAFATOS
[1]*Center for Earth Observing and Space Research*
[1]*Institute for Computational Sciences and Informatics and* [2]*Department of Physics, George Mason University, Fairfax, VA 22030, USA*

1. INTRODUCTION

The recent experiments on long-distance Bell-type tests[1] clearly indicate that quantum correlations between photons can persist for very long distances at least over 10 km. These experimental results support the view that distance has no effect on quantum correlations. We propose that the cosmological scale might be an arena to test those long distance quantum correlations.

Advances in the theory of elementary particles have afforded us a unique opportunity to describe theoretically the conditions of the universe close to the big bang. It is agreed that the early universe was in a quantum state and that modeling the early universe can be used as a cosmic laboratory to push physics to the furthest frontiers that it has ventured to go, namely to the limits of quantum gravity[2]. Progress in cosmology in the last few years has been quite stunning due to largely the success of the quantum field theory of particle interactions. The evolution of the cosmos as is presently understood in big-bang theory is outlined here.

Regarding the beginning of the universe, quantum gravity will likely provide clues as to the actual origin of the cosmos. In the first phase of evolution after the original singularity the space-time description breaks down entirely, and this is followed by the so called "inflationary era" at about 10^{-35} sec[3]. During the inflationary era, the universe underwent an extremely rapid expansion doubling in size every 10^{-35} of a second. By the time the universe had undergone inflation, it had expanded in size by a

Studies on the Structure of Time: From Physics to Psycho(path)ology
Edited by Buccheri *et al.*, Kluwer Academic / Plenum Publishers, New York, 2000.

staggering factor of 10^{50} or more. The inflation was a phase transition, similar to the phenomenon of freezing of water. In fact, the big bang model with inflation speculates that the universe first came into existence as a result of a quantum transition. After inflation had released the vast amounts of energy that would later on coalesce to form all the observable objects in the universe, the expansion proceeded according to the original version of the big-bang theory. Between 10^{-30} sec and 10^{-6} sec the universe was filled with a primordial soup of quarks and light particles like the electrons. After about 10^{-6} sec from the big bang, the quarks combined together to form the heavy particles like the protons and the neutrons. In the next phase of ascending complexity, between 1 sec and 3 minutes, protons and neutrons underwent nuclear reactions forming nuclei of helium and its isotopes. Although some heavy elements were formed in this period as well, their amounts were minute and the formation of appreciable amounts of those elements occurred only many millions of years later in massive stars that evolved into exploding supernovae, hurling the matter processed by nuclear reactions in their centers into space.

Up until about 100,000 years after the big-bang photons and matter were coupled together and this means that the universe was dark, or light could not travel very far. The $2.73K$ black body photons which are one of the primary means of supporting big-bang cosmology were also emitted at this time but provided no opportunity following this era to probe the earlier periods where fundamentally important physical processes were taking place. However, the cosmic microwave background radiation exhibits a remarkable degree of homogeneity. It has the same temperature to within one part in ten thousand or more in every direction on the sky. Observations to detect intrinsic (not due to the motion of the observer) inhomogeneities at various angular scales have only resulted in upper bounds on $(\Delta T/T)$[4]. This homogeneity of microwave blackbody radiation in hot big-bang cosmology prevails even though this radiation was formed 10^5 years from the beginning, when opposite sides of the sky at that time were out of causal contact, separated by $\sim 10^7$ light years. Given the virtual identity of temperatures from all parts of the sky and presuming that classical physics holds, one would conclude that opposite parts of the sky had to be in causal contact. The theory of relativity, though, assumes that no signal can travel faster than light, and opposite parts of the sky were space-like in separation. This is known as the horizon problem in cosmology usually explained away by the various inflationary scenarios. It represents a problem of incredibly fine tuning in the conditions prevailing in the early universe or else a violation of classical causality. Other correlations in the large-scale structures of the universe exist such as very large structures in the distribution of matter[5] which may continue in progressively larger and larger clusters of matter all the way to the scale of the universe itself.

Physicists like Barrow and Tipler[6] proposed the Anthropic Principle to account for the above fine tuning properties of the universe. They have postulated that our existence as observers requires this fine tuning. On the other hand, most cosmologists tend to view the universe as an idealized closed system that is isolated from the observer and his observing apparatus. Here, we shall consider that the universe is more like a quantum rather than a classical system. Quanta in the early universe were tangled together and one expects a similar situation as in the Bell-type experiments.

Here, as an alternative to inflationary scenarios or considerations such as the Anthropic Principle, we postulate the remarkable correlations exhibited at the cosmological scales which are considered to be reminiscent of Bell-type quantum correlations. One of the authors (Kafatos) proposed that Bell-type correlations[2] would be pervasive in the early universe. Binary scattering processes involving the resulting outcomes of these processes (even in the absence of observer) would produce space-like correlations among the original entangled photons. It is worth mentioning that Padmanavan and Narlikar[7] have proposed similar ideas in order to explain the homogeneity of CMBR by invoking the following quantum mechanical features:

1. An observation performed on a quantum system 'collapses' the wave function of the system instantaneously all over a space-like surface;
2. Such an irreversible act of observation, however, need not occur on that particular hypersurface; in other words, the choice as to the nature of the measurement may be delayed, taking place billions of years later.
3. The above aspects can lead to correlations between results of measurements performed at events which are not causally connected.

Based on these quantum mechanical considerations they tried to show that the homogeneity of CMBR may be viewed as a consequence of a causal correlation between results of measurements of CMBR. In fact they extended this argument so that it can be used in a cosmological context as follows: In the conventional big bang cosmology, the temperature of the CMBR is taken as an indicator of the physical state. It immediately follows that CMBR temperatures should appear correlated over space-like hypersurfaces. As the temperature can be related to the energy of the photon in CMBR we can think of the correlation of frequencies of the photons. It is well known that an observation on a quantum system collapses the wave function of the system instantaneously all over the space-like surface. Such an irreversible act of observation, however, need not occur on that particular hypersurface; in other words, the choice as to the nature of the measurement may be delayed, taking place billions of years later. This aspect can lead to correlations between results of measurements performed at events which are not causally connected. In this kind of delayed choice experiment, temporal non-locality occurs where the path that a photon follows is not determined until a delayed choice is made. Now at cosmological scales, one can think of the existence of Bell-type correltions using energy-time entangled photons as performed by Tittel et al[1]. In this paper, we shall consider the binary scattering processes in section 2 and temporal non-locality in section 3.

2. BELL-TYPE CORRELATIONS & LARGE SCALE STRUCTURE

We simply argue here that the universe is more like a quantum rather than a classical system. It certainly can not be viewed as a closed system in the sense of being separate from the observer. Quanta in the early universe were tangled together which can be visualized as follows:
In the electron-positron annihilations, the resultant gamma rays would be polarized perpendicular to each other i.e,

$$e+e^{-}\rightarrow\gamma+\gamma'$$

$$\gamma+e^{\pm}\rightarrow\gamma'+e^{\pm}$$

etc. Thess type of binary processes dominates as $z\rightarrow 1{,}000$ or near the matter-radiation decoupling era.

Now since Compton scattering strongly depends on polarized radiation, once one of the photons gets scattered, in a subsequent process an 'experimental choice' has been made, even in the absence of an observing apparatus. The very physical process of scattering provides the choice. The other photon's polarization would not then be presumed to be independent. It would have to be a well defined polarization and for the subsequent scattering of the second photon, an isotropic distribution of polarization could not be assumed.
Let the first photon be polarized at an angle θ_1 and the second at θ_2. Then following the subsequent scatterings we can get successfully higher orders of scattering cascades, first as $\theta_1, \theta_1', \ldots, \theta_1''$ and the second order as $\theta_2, \theta_2', \ldots, \theta_2''$ Now each array of scattering can be thought of as a sequence of measurements as if n-polarizers are placed sequentially at angles $\theta_1, \theta_1', \ldots, \theta_1''$ and similarly for θ_2. Again during these sequential measurements the external noise-like primordial density fluctuations may

produce degradation in the correlation or decoherence (work under preparation). Then we need to consider N(γ+e) Compton scattering process. The two particle correlation and inequalities have been generalized to n-spin particles by Mermin[8]. Consequently we can think of Bell-type correlations among the polarized photons in terms of the frequencies of the photon instead of directly in terms of temperature. Moreover, it has been shown[9] that two photon correlations remain strong enough over distances of more than 10 km. We can calculate the differential crossection for compton scattering with polarized photons using the correct Klein-Nishina formula. It is worth mentioning that this kind of correlations attracted a lot of attention after the pioneering work of Bell. "Typical theoreticians" took Bell's work as a "Calculus of Chimeras" i.e. if there existed hidden variables, the integration over them would result in some (incorrect!) inequalities. Does it pay to master the art of estimating integrands over hidden variables which do not and can not exist? Then the quantum analog of Bell's inequalities was obtained by an operator technique that is far from "chimera". The classical Bell inequalities can be considered to apply for commuting operators. They may be applied to the factorized states and their mixtures. Thus, after Bell's inequalities had played a historic role in rejecting local hidden variables, they found a second life within the quantum theory in the "crystallization centers" of a quantum correlation[10]. The meaning of this correlation function can be understood in the following manner. Quantum theory interwines mutually exclusive (complementary) experimental arrangements and leads to general restrictions on complementary experimental data which do not depend on the specific representation of observables within the theory. Then one can work out the details for the EPR-type experimental set up, or the quantum restrictions on empirical two point correlation functions.

The kinematics of correlations as treated in the framework of axiomatic local quantum field theory, was born in 1985. In itself it is not new. But the results are independent of the notions and methods specific to field theory, such as Lorentz invariance, vacuum state, spectrality conditions, mass gap, Ruch-Schleider theorem etc. The kinematics of correlations embedded in the free field theory is distinguished by the use of canonical commutation or anticommutation relations for the field operators. Summers and Werner[11] have shown that the correlation function $R \approx 2\sqrt{2}$ in the vacuum state if proper observables are selected from local algebras of free boson or fermion fields (massive or massless) and two regions in the regions of space-time are selected as mutually complementary wedges. The vacuum state of the free boson field itself proved to be similar to the state considered by Einstein, Podolsky and Rosen. In this kinematics of correlations, we should emphasize that one utilizes several commuting algebras which correspond to several space-like separated regions in space-time. But it is necessary to develop

further and more substantive multiple-time theory i.e, describing dynamics of correlations for which the above kinematics might be a prologue.
However, the above correlation functions may be useful to help us to understand the large scale correlations of the present universe. After the last scattering epoch, as the expansion of the universe was continued, the correlation functions connecting the observables for photon frequencies (Bell-type correlation) would be expected to be prevalent as the general kinematics is valid for separated space-like regions. The role of evolution of the universe may be explicitly considered in the correlation functions when it would be possible to develop a complete dynamical theory of correlations. Here we postulate that the characteristic length associated with this kind of correlation function can be related to the density of the universe, which itself is changing as the universe is expanding i.e. the correlation length is changing in the expanding universe.

3. TEMPORAL NON-LOCALITY

It is clear from the above analysis that the large scale structure of the universe can be related to Bell-type correlations for the frequencies of CMBR photons in the early universe. This non-local correlation is known as temporal non-locality.

Kafatos and Nadeau[12] discussed three types of non-locality.

Spatial or Type I non-locality occurs when 2 quanta (such as photons) remain entangled at all scales across space-like separated regions, even over cosmological scales.

Temporal or Type II non-locality (or Wheeler's delayed choice experiment[13]) occurs in situations where the path that the photon follows is not determined until a delayed choice is made.

Type III non-locality represents the unified whole of space-time revealed in its complementarity aspects as the unity of space(Type I) and the unity of time (Type II). It exists outside the framework of space and time and cannot, therefore, be discerned by the scientific method although its existence is implied.

Here, we have a quantum non-local correlation prevalent throughout the history of the universe. One can consider an experimental situation which deals with the largest system imaginable. However, one cannot in theory at least even presume a categorical distinction between acts of observation and the observed system even if the system is in excess of the billion galaxies in the observable universe. In the absence of the observing apparatus, as the "choice" is made by the very act of physical processes , is a situation very similar to Wheeler delayed choice type of experiments. So the type II non-locality can be thought to be prevalent in the cosmological domain. So far the type II non-locality has been tested in the laboratory scale. We would like to emphasize that this kind of temporal non-locality may play significant role in explaining the large-scale structure and homogeneity problems of the universe.

REFERENCES

1. Tittel W., Brendel J., Gisin N., and Zbinden H., 1999, "Long-distance Bell-type test using energy-time entangled photons", *Phys. Rev. A.*, **59** 4150.
2. Kafatos M., 1986, "Horizons of Knowledge in Cosmology" in Bell's Theorem, *Quantum Theory and Conceptions of the Universe*, Kluwer Academic Publishers, Amsterdam, 195.
3. Guth A. and Steinhardt P.J., 1984, "Inflationary Universe", *Scientific American*, May.
4. Smoot G.F., 1984, *Examining the Big Bang and Diffuse Background Radiation*, Kafatos and Kondo (eds.), Dordrecht, Kluwer Publishers.
5. Geller M.J. and Huchra J., 1989, "Mapping of the Universe", *Science* 246, 897.
6. Barrow J.D. and Tipler F.J., 1986, *The Anthropic Cosmological Principle*, Oxford University Press.
7. Padmanabhan T. and Narlikar J.V., 1998, Homogeneity of the Cosmic Background Radiation as Consequences of a Quantum Delayed Choice Experiment, private communication.
8. Mermin D.N., 1990, "Extreme Quantum Entanglement in a Superposition of Macroscopically Distinct States", *Phys Rev. Lett.*, **65** 1838.
9. Tittel W., Brendel J., Zbinden H., and Gisin N., 1998, "Violation of Bell Inequalities by Photons more than 10 km apart", *Phys. Rev. Lett.*, **81** 3563.
10. Khalfin L.A. and Tsirelsm B.S., 1992, "Quantum/Classical Correspondence in the light of Bell's Inequalities", *Found. Phys.* **22** 879.
11. Summers S. and Werner R., 1987, "Bell's Inequalities and Quantum Field Theory I", *J. Math. Phys.*, **28** 2440.
12. Kafatos M. and Nadaeu R., 1990, *The Conscious Universe*, Springer-Verlag.
13. Wheeler J.A., 1981, *Some Strangeness in the Proportion*, Woolf (ed.), Addison-Wesley, Reading, Mass.

Algebraic Geometry: a Tool for Resolving the Enigma of Time?

METOD SANIGA
Astronomical Institute, Slovak Academy of Sciences, SK-059 60 Tatranská Lomnica, The Slovak Republic

1. INTRODUCTION

Although time is a concept that attracted and occupied the thoughts of a countless number of thinkers and scholars over centuries, its true nature still remains wrapped in a shroud of mystery. Perhaps the main reason why this is so is that there are too many aspects this notion entails. Out of these, there are two that deserve particular attention for, in our opinion, they characterize best the nature of time's enigma. The first is the notion of *physical* or *objective* time. Time, from the physicist's viewpoint, is one of the dimensions of a differentiable, Riemannian manifold. It is a linear variable, structureless and undifferentiated. It is very much like a spatial dimension, the only difference between the two being embodied in the Lorentz signature of the metric tensor. The second is *psychological* or *subjective* time; time as we perceive it, experience it. This time dimension exhibits a non-trivial internal structure, consisting of the past (events that have happened) and the future (events that will happen), the two domains being separated from each other by a unique moment of the present, the event that is happening. This is the time that seems to "flow", to proceed at a steady pace from the past into the future - the time with an arrow pointing in one direction.

It is a well-known fact that the fundamental equations of physics are time-reversible, i.e. they do not distinguish between the past and future. Moreover, the very concept of the present, the now, has no proper place in the temporal of physics at all; this holds true whether one is talking of classical physics, quantum mechanics or relativity theory. This is because

Studies on the Structure of Time: From Physics to Psycho(path)ology
Edited by Buccheri *et al.*, Kluwer Academic / Plenum Publishers, New York, 2000.

physics merely *quantifies* the points of the time dimension, thus stripping the latter of its crucial experiential aspect - its arrow. The physical time is thus a plain, quantitative mathematical abstraction and, as such, it simply cannot account for the fine structure of the subjective time. In order to capture the latter, it is, therefore, necessary to go beyond quantifying and consider a generalized mathematical concept of dimension where a *qualitative* distinction between the individual elements (or, at least, between the groups of the latter) can be implemented.

An illustrative example of such a temporal dimension is provided by a specific linear, single-parametric set (so-called pencil) of conics in the projective plane[1,3–8]. This set of conics is found to nicely reproduce the experienced arrow of time when the projective plane is affinized; it simply suffices to postulate that each proper conic of the pencil stands for a single temporal event, and relate three distinct kinds of (proper) affine conic, viz. a hyperbola, a parabola and an ellipse, with the three different kinds of temporal event, viz. the past, present and future, respectively[1–3]. The model, albeit very simple, turns out to be very useful and fertile, for in addition to the ordinary arrow, it also gives rise to a number of other, differently structured features of the temporal. These patterns emerge as a combination of two important properties of the projective plane: the *way* in which the plane is affinized and the *character* of its ground field[4–8]. And the fine structure of psychological time is, indeed, found much richer and more intricate than outlined above[9–136].

Already in a normal state of health there are, every now and then, aberrations of subjective time such as acceleration or deceleration of the lapse of time. Under severe mental disturbances (like those characterizing mental psychoses, drug-induced states, trances, mystical states as well as other deep "altered" states of consciousness), these anomalies/peculiarities become more pronounced. The flux of time may even cease completely (the sensations usually described as "time standing still", or "suspended", "arrested" time), or expand without limit (the feelings of "everlasting now", "eternity"). In some cases, time's flow may be experienced as discontinuous, fragmented, or even reversing its direction. Finally, in most extreme cases, time as a dimension is transcended, or simply non-existent (an "atemporal" state).

It is remarkable that already our pencil-model is able to grasp, at the semi-quantitative level, many of the above-introduced "distortions" of the subjective experience of time[4–8]. There are a couple of serious implications this fact conveys to us. On the one hand, we see that although the manifestations of psychological time are so diverse, unusual and failing to conform to any current physical paradigm, they are accommodated by a definitive algebraic geometrical pattern. On the other hand, we are inevitably led to the murky, bizarre and often contra-intuitive realm of phenomenology, which has always been potentially dangerous to enter. Yet this is a territory whose "topology" is very likely to contain a decisive clue to ultimate unveiling of the nature of time - all the more that the methods and results are in no case less mathematically rigorous than the currently widely-accepted practice of dynamic modeling of biological systems.

There is, however, another important message the above-listed examples of the anomalies/pathologies in time perception try to impart on us. It is the fact that the intrinsic structure of the arrow, an intriguing aspect of which is the pace at which time seems to "flow" is not rigid, but rather flexible and subject to profound changes and dramatic alterations. This means that the subjective time has a strongly *non*-linear character. As a consequence, we have to abandon its pencil-model and consider a more general single-parametric aggregate of conics as its representative.

2. A SIMPLE QUADRATIC GENERALIZATION OF THE PENCIL-MODEL

It is obvious and natural that we demand this non-linear aggregate of conics to be the simplest possible generalization of the pencil of conics employed in our debut model[1,3]. The pencil in question enjoys three important properties. Firstly, it features linear base (i.e. shared by all the conics) elements*, i.e., its base elements are either points or lines. Secondly, it is self-dual, i.e., it looks the same irrespective of the representation adopted. Finally, it is full, i.e., all its base elements are real, or, in the case of a general ground field, all belonging to the field. Hence, we are to seek for such an aggregate that also meets these constraints.

To begin with, we recall the *duality* principle of the projective plane, i.e. the fact that the projective geometry of the plane exhibits a twofold character. Namely, the set of all the points of the plane and the set of all the lines of the plane are symmetrically related to each other, and to every property of lines in the geometry of points there corresponds a property of points in the geometry of lines. There is thus a perfect correspondence between the algebra of line-geometry and that of point-geometry; we say that a point and a line of the projective plane are dual to each other entities. In other words, instead of viewing the points of the plane as the fundamental entities, and the lines as the ranges (loci) of points, we may equally well take the lines as the primary entities and define points in terms of lines, characterizing a point by the complete set of lines passing through it. The

* The terminology, symbols, and notation used here are the same as in ref. 3-8.

points and lines of the projective plane may thus be regarded as two equivalent representations of the latter; a(ny) projective theorem remains valid if the concepts of point and line are interchanged. A given geometrical object of the plane will, in general, look different in the two representations of the plane; if it looks the same, it is called self-dual. So, if our aggregate of conics is to be self-dual, there must be parity between the number of its base points and that of its base lines.

Next, a proper point-conic $\mathcal{Q}$,

$$\mathcal{Q}(c) \equiv \sum_{i,j=1}^{3} c_{ij}\breve{x}_i\breve{x}_j \equiv$$

$$\equiv c_{11}\breve{x}_1^2 + c_{22}\breve{x}_2^2 + c_{33}\breve{x}_3^2 + 2c_{12}\breve{x}_1\breve{x}_2 + 2c_{13}\breve{x}_1\breve{x}_3 + 2c_{23}\breve{x}_2\breve{x}_3 = 0\,, \quad (1)$$

$$\det c_{ij} \neq 0\,, \quad (2)$$

is defined by six distinct quantities c_{ij}'s. Of these, however, only five are independent for $c_{ij} = \rho c_{ij}$ ($\rho \neq 0$) clearly represent the same conic. In order to get a single-parametric set of conics, we have thus to impose four different constraints on the values of c_{ij}'s. Making the conic pass through a given point is a simple constraint; dually, requiring the conic to touch a particular line is also such a constraint. Hence, a self-dual, single-parametric aggregate of conics with linear base elements must feature *two* base points and *two* base lines.

In the case of our favorable pencil of conics, the base points (B and B') lie on the base lines (β and β', respectively) as shown in Fig. 1a [1,3]. It turns out that in order to get a simplest non-linear aggregate of conics with the desired symmetry properties, it suffices to relax this incidence relation for one point-line pair - see Fig. 1b.

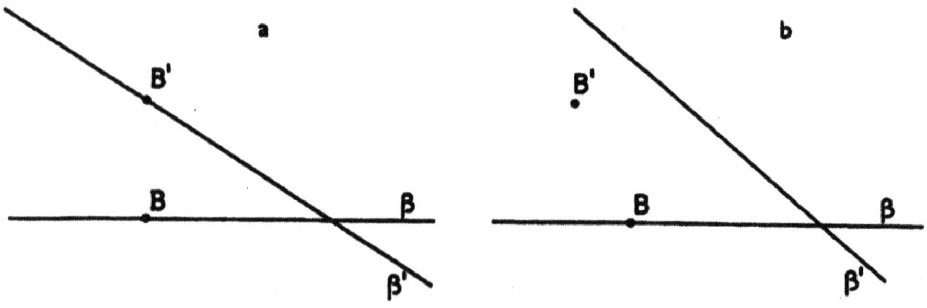

Figure 1. The configuration of the base elements of our favorable pencil of conics (*a*) and of (one of) its simplest non-linear generalization(s) (*b*).

Let us find the equation of a generic conic of this aggregate. In order to keep reasoning at the simplest possible level, we choose the coordinate triangle in such a way that the base points B and B' lie on its $\breve{x}_3=0$ side and coincide, respectively, with its vertices V_2 and V_1, that is

$$B:\ \rho\breve{x}_i = (0,1,0),\ \rho{\neq}0, \tag{3}$$
$$B':\ \rho\breve{x}_i = (1,0,0),\ \rho{\neq}0, \tag{4}$$

and the base lines are selected as follows

$$\beta:\ \breve{x}_1 = 0, \tag{5}$$

and

$$\beta':\ \breve{x}_1 = \kappa\breve{x}_2,\ \kappa{\neq}0, \tag{6}$$

as illustrated in Fig. 2.

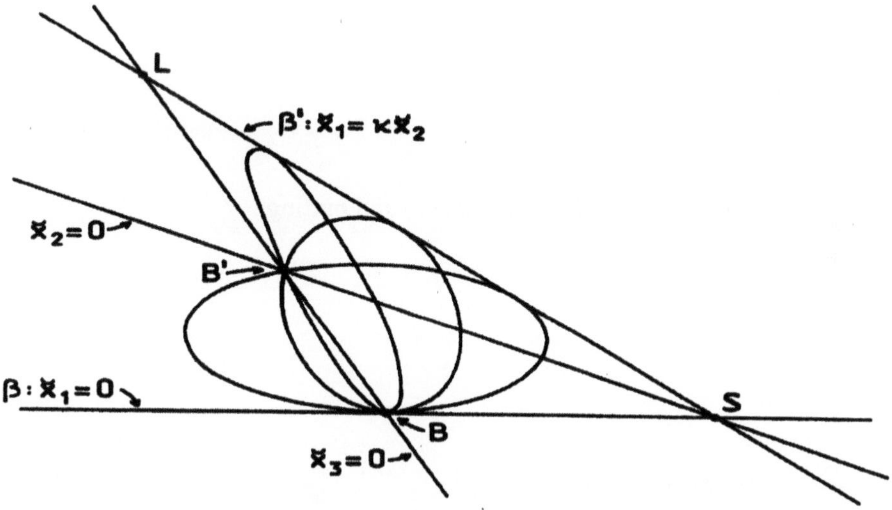

Figure 2. Selection of the coordinate system which facilitates description and examination of the non-linear aggregate of conics with the base configuration as depicted in Fig. 1b. Also shown are three proper conics of the aggregate, belonging to the same family.

Now, a generic conic of the projective plane, $\mathcal{Q}$, is given by eqn. (1). If this conic passes through B and B', then

$$c_{22}=0 \tag{7}$$

and

$$c_{11}=0, \tag{8}$$

respectively. Further, β is a tangent to $\mathcal{Q}$ iff

$$c_{23}=0; \tag{9}$$

for substituting eqn. (5) into eqn. (1) and taking into account eqn. (7) gives

$$c_{33}\breve{x}_3^2 + 2c_{23}\breve{x}_2\breve{x}_3 = 0, \tag{10}$$

which indeed yields repeated roots (in the variable $x \equiv \breve{x}_3/\breve{x}_2$) under eqn. (9). Similarly, inserting eqn. (6) into eqn. (1) and taking into account eqns. (7)-(9) leads to

$$c_{33}\breve{x}_3^2 + 2\kappa c_{13}\breve{x}_2\breve{x}_3 + 2\kappa c_{12}\breve{x}_2^2 = 0, \tag{11}$$

from where it follows that β' touches $\mathcal{Q}$ iff

$$c_{12} = \frac{\kappa c_{13}^2}{2c_{33}}. \tag{12}$$

Combining eqns. (7)-(9), (12) with eqn. (1) and putting

$$\vartheta \equiv \frac{c_{13}}{c_{33}} \tag{13}$$

we thus find

$$\mathcal{Q}(\vartheta) = \kappa\vartheta^2\breve{x}_1\breve{x}_2 + 2\vartheta\breve{x}_1\breve{x}_3 + \breve{x}_3^2 = 0. \tag{14}$$

We see that the aggregate is indeed a non-linear one, for the last equation depends *quadratically* on the variable parameter ϑ. It contains two degenerate conics,

$$\mathcal{Q}(\vartheta = 0) = \breve{x}_3^2 = 0 \tag{15}$$

and

$$\mathcal{Q}(\vartheta = \pm\infty) = \breve{x}_1\breve{x}_2 = 0\,, \tag{16}$$

which separate the set of the proper conics into two distinct families, viz. $-\infty<\vartheta<0$ and $0<\vartheta<+\infty$.

3. VARIETY AND STRUCTURE OF THE AGGREGATE-BORNE TIME DIMENSIONS

Our principal task is to find all the possible types of temporal arrow borne by this aggregate of conics and examine in detail the intrinsic structure of each of them. The strategy to be followed is fully paralleled that pursued in refs. 1-8. Namely, either family of aggregate (14) is taken to represent a single time dimension, every proper conic being viewed as a particular event. We thus have a pair of time dimensions, which can be regarded as twin dimensions. They are both originally homogeneous/structureless, because from the projective point of view all proper conics stand on the equal footing. To 'de-homogenize' them, we pick up one line of the projective plane (henceforth referred to as the *d*-line - broken line in Figs. 3 and 4) to play a special role, and examine intersection properties of this line with the conics of the aggregate. The (proper) conics that are cut by this line in two distinct points (solid curves in Figs. 3 and 4) are then postulated to stand for the past events, those having no points in common with it (dotted) are regarded to represent the future events, and the unique conics featuring this line as a tangent (dot-dashed) are identified with the moments of the present.

Due to a quadratic nature of the aggregate, there are, in general, *two* distinct conics which are touched by the *d*-line. They can be found to belong to different families (Figs. 3a, 4a; 3g, 4g), or lie in the same family (Figs. 3c, e). This can easily be substantiated quantitatively. We define *d*-line by the equation

$$\breve{x}_2 = a\breve{x}_1 + b\breve{x}_3\,, \tag{17}$$

a and b being fixed real parameters, and combine this equation with eqn. (14) to obtain

$$x^2 + \vartheta(b\kappa\vartheta + 2)x + a\kappa\vartheta^2 = 0, \quad x \equiv \breve{x}_3/\breve{x}_1\,. \tag{18}$$

Obviously, *d*-line is a tangent to those conics of (14) whose ϑ's lead to coinciding (repeated) roots of eqn. (18),

$$\vartheta^2\left[(b\kappa\vartheta+2)^2-4a\kappa\right]=0\,. \tag{19}$$

The last equation is satisfied by ϑ=0, which is not a proper conic (see eqn. (15)), as well as by

$$\vartheta_{\pm}=\frac{-2\pm2\sqrt{a\kappa}}{b\kappa}\,. \tag{20}$$

Hence, if $a\kappa$ is positive and less than one, ϑ_+ and ϑ_- are of the same sign, i.e., the conics are in the same family; on the other hand, for $a\kappa$ exceeding one ϑ_+ and ϑ_- have the opposite signs and, hence, the two conics belong to different families. If $a\kappa$=1, there is only one proper conic enjoying this property (that corresponding to ϑ_-; the other conic, ϑ_+=0, is degenerate), as it is also the case when $a\kappa$=0 ($\vartheta_+=\vartheta_-=-2/b\kappa$). Finally, for negative κ there are no such conics. Rephrased in terms of physics/psychology, there are two present moments for positive $a\kappa$, which can be either situated in the same time dimension ($a\kappa$>1), or shared by the two temporal coordinates ($a\kappa$<1), a single-now time dimension for $a\kappa$=0, 1, and the presentless temporal for $a\kappa$<0. As it is only the present, the 'now', that is directly perceived, it is natural to further stipulate that out of the two potential dimensions only that is observable which contains a present moment; the one lacking this particular moment is viewed as latent, unobservable. From what has just been said it follows that if the position of the *d*-line with respect to the aggregate varies, the corresponding time configurations will differ not only in their internal structure, but also in their dimensionality. This is the most crucial aspect of our theory and we shall, therefore, have a detailed qualitative look at it.

We shall begin with the case when both the families exhibit a *perfect arrow*, Figs. 3a, 4a. In order to see the link between different time patterns most explicitly, we shall move the *d*-line in such a way that it remains all the time tangent to one and the same conic of a given family - the family shown in Fig. 3, the conic in question drawn dot-dashed. The shift is done in a manner that the point of contact moves on the conic counter-clockwise. Starting, as mentioned above, from a perfect-arrow pattern, this structure is qualitatively preserved till *d*-line gets identical with β' - Figs. 3b, 4b. In this case the line is a tangent to all the proper conics of both families, and so both time dimensions comprise solely the present moments; in what follows we shall call this type of the temporal the *pure present of the first kind.*

Shifting *d*-line further (Figs. 3c, 4c), we arrive at a mode where there is a profound distinction in the structure of the two dimensions. While one dimension (Fig. 4c) becomes homogeneous (and, hence, unobservable), built exclusively of the past events, the other (Fig. 3c) is structured in a highly non-trivial way: namely, it features two distinct domains of the past that are separated by two different moments of the present from a single region of the future. This can be viewed as a union of two ordinary arrows of time that have the domain of the future in common. Although the two arrows are perfectly equivalent on their own, this symmetry is broken when this structure (Fig. 3c) is related/compared to that depicted in Fig. 3a. For from this comparison it follows that one of the arrows, viz. the one that shares with the arrow of Fig. 3a the moment of the present, can be regarded as principal (illustrated in Fig. 3c by the double arrow); the other, seen located in what would normally be the region of the future of the principal arrow, as subsidiary (single arrow). We note that the principal arrow is opposite to the 'reference' arrow of Fig. 3a, whereas the subsidiary one points in the same direction as the latter. In what follows we shall call this configuration an *arrow within arrow*.

This structure remains qualitatively intact until *d*-line is found to pass through the base point B', Fig. 3d. In this case, the domain of the future completely disappears and the time dimension is endowed with two regions of the past, separated from each other by a single moment of the present: this pattern will be referred to as a *past-present-past* mode. Concerning the twin time dimension, it still keeps its latency as the corresponding family of conics preserves its homogeneity (Fig. 4d).

A further slide of *d*-line brings us again to an arrow-within-arrow configuration, Fig. 3e. This a-w-a pattern is, however, related differently to the original arrow than that represented by Fig. 3c. For it is now the domain of the future of the reference arrow (Fig. 3a) that is transformed into a subsidiary arrow, and it is the principal arrow which shares the polarity of the reference arrow. In order to distinguish between the two a-w-a configurations, we shall call the former (Fig. 3c) the *past*-related, the latter (Fig. 3e) the *future*-related one. The corresponding second temporal dimension is again hidden, as it still features past events only (Fig. 4e).

The a-w-a pattern of Fig. 3e vanishes when *d*-line hits the point L, and a perfect arrow is re-established - Fig. 3f. This pattern is seen all the way through to the point when *d*-line becomes identical with the other base line, β - Fig. 3h. At this mode, all the conics have *d*-line for a tangent and we again have a time dimension consisting exclusively of the 'nows', the present; as it will be seen later on, this mode is fundamentally different from that represented by Fig. 3b, and to make this distinction evident we shall call it the *pure present of the 2nd kind*. This is the last pattern in the sequence, for a further displacement of *d*-line leads to the perfect arrow we began with (Fig. 3a, 4a).

Summing up, we arrive at six different patterns of the intrinsic structure of the time dimension borne by one family of the quadratic set of conics

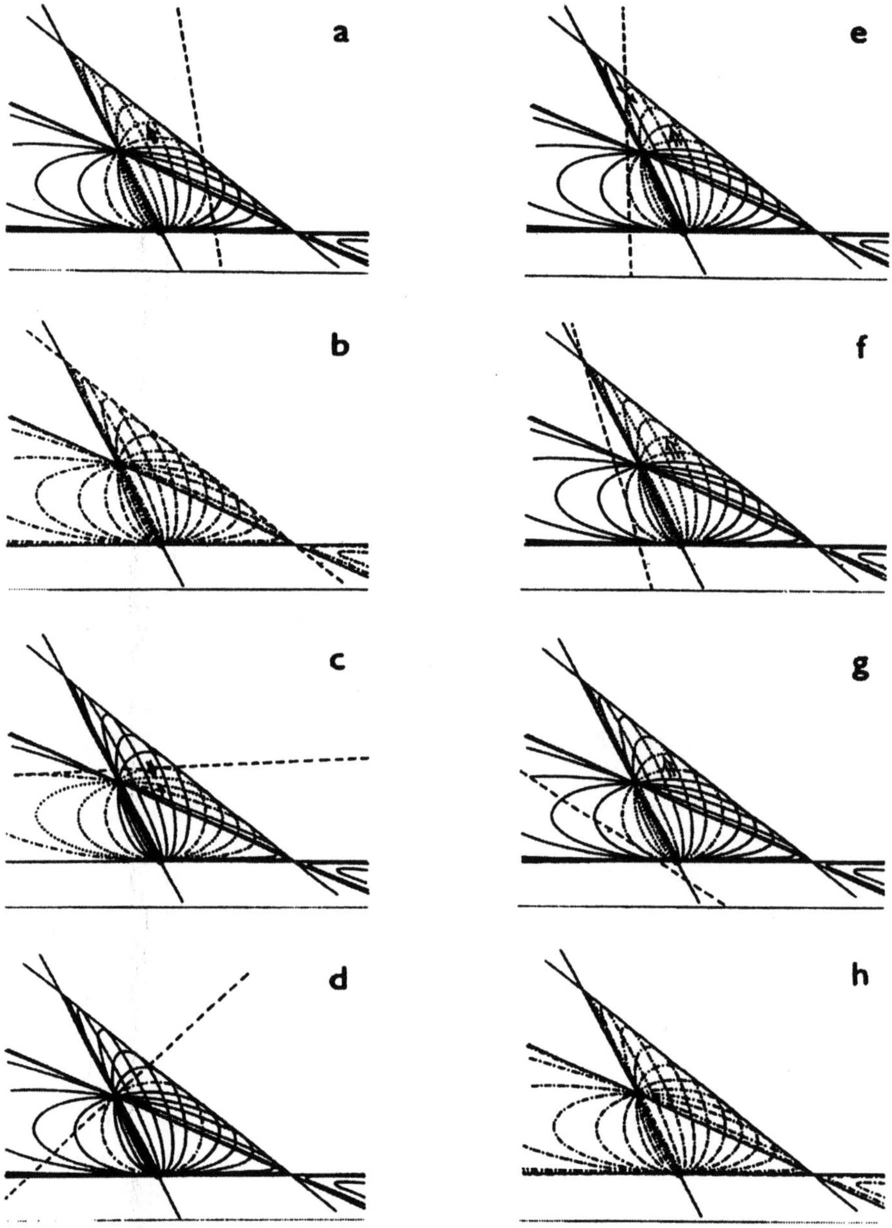

Figure 3. A schematic sketch of the varieties of the internal structure of time dimension generated by a given family of conics of the aggregate defined by eqn. (14): *a* - a perfect (ordinary) arrow; *b* - a pure present of the 1st kind; *c* - a past-related arrow within arrow; *d* - a past-present-past pattern; *e* - a future-related arrow-within-arrow mode; *f*, *g* - perfect arrows, and *h* - a pure present of the 2nd kind. For more details, see the text.

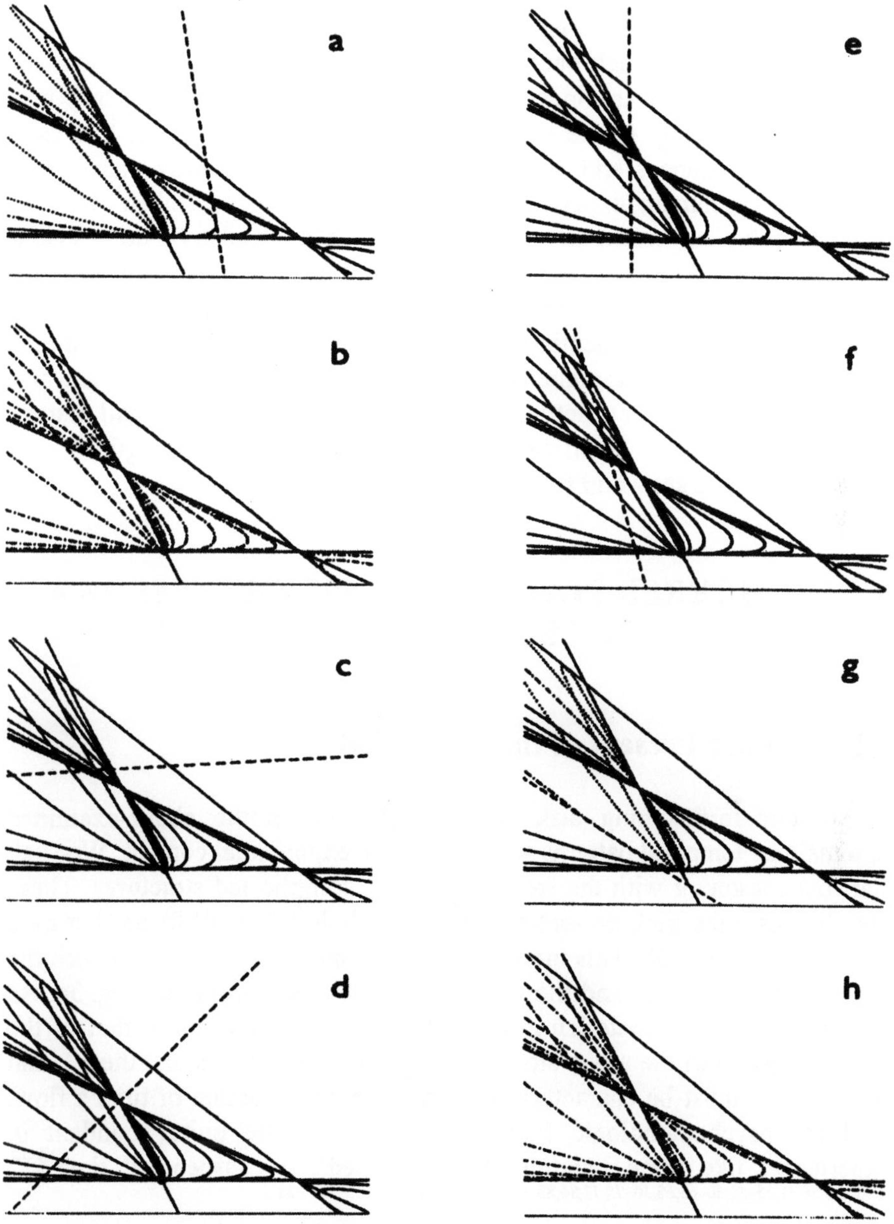

Figure 4. The same as in the previous figure for the time dimension generated by the other family of conics of aggregate (14). The position of *d*-line on the corresponding figures is identical.

defined by eqn. (14): the perfect arrow (Figs. 3a, f, g), a couple of pure-present configurations (Figs. 3b, h), two arrow-within-arrow modes (Figs. 3c, e), and a past-present-past structure (Fig. 3d). The perfect arrow of time is a structure which each of us is familiar with; but what about the remaining five? They are, indeed, quite peculiar and, at first sight, they may seem to belong only to the realm of pure academic interest. In this case, however, appearances and reality differ perhaps more greatly than in any other area of human experience, for the theory can be linked with the phenomenological world in a way that has had absolutely no previous existence of any kind, of which science has not allowed us to form even the remotest conception. What the term 'phenomenological' world means here is nothing but the already-mentioned (Sec. 1) realm of psychopathological time dimensions characterizing altered states of consciousness[9–136]. It is a world where there is so much unsettled and uncertain that seemingly small matters can take an outsize significance and where, thus, the strength of our formalism can be shown in full.

4. ALTERED STATES OF CONSCIOUSNESS AS A VALUABLE TEST BED OF THE MODEL

4.1 Pure Present of the First Kind

So, our forthcoming task is to try to fill up the above-examined 'anomalous' temporal patterns with a concrete experiential content. We shall start our reasoning with the (relatively) least complicated structures. These are obviously the pure-present modes, Figs. 3b, h. We shall focus first on a 1st kind mode, Fig. 3b. This mode represents a smooth transition between the perfect arrow, Fig. 3a, and the past-related a-w-a configuration, Fig. 3c. As the flux of time characterizing the perfect arrow is opposite to that of the principal arrow of the a-w-a structure in question, it follows that the 1st kind pure present must be characterized by a complete cessation of time's flow. And the psychopathologic literature is found to be quite abundant in experiences like 'time standing still', 'arrested', or 'suspended' time[10–26, 28–30,36,38,41,45–52,60–66,68,70–75,77,84,88–93,102,106–108,113,118–129,132].

A quadruple of descriptions given below were drawn from different sources, and they should give the reader a sufficiently complex picture of what a generic experience of this type of disturbed temporal awareness looks like. The first excerpt was taken from a beautiful paper by Muscatello and Giovanardi Rossi (ref. 123, p. 784)*:

"*Il tempo per me è fermo completamente*, mi sembra. Forse sono soltanto attimi che io sto così male. Guardo l'orologio e mi sembra, se lo riguardo, che sia trascorsa una distanza di *tempo enorme*, come se fossero trascorse delle ore, mentre invece sono trascorsi pochi minuti. Mi sembra un *tempo enorme* che sta passando. *Il tempo non passa mai, guardo l'orologio, ma le lancette sono sempre ferme nella stessa posizione, non si muovono mai, non vanno mai avanti; allora sento se l'orologio si è fermato, sento che batte, ma le lancette sono sempre ferme*. Al passato non penso, me ne ricordo ma non ci penso molto. Quando sto così male *non penso mai al mio passato*. Non mi veniva in mente niente, niente... non riuscivo a pensare a niente. Non riuscivo a prospettarmi niente, *niente in futuro. Per me il tempo presente non esiste* quando sto così male... *non esiste il passato, non esiste il futuro*."

An extraordinary vivid and detailed account of this type of the psychopathology of time was found in ref. 77 (p. 311):

"...Das seltsame war, daß mir hier und da das normale Zeitbewußtsein angesichts dieser Figuren vollkommen verloren ging; *die Zeit war nich mehr ein Strom, der vorbeifloß und dessen Fließen man messen konnte, sondern sie war gleichsam ein Meer, das als Ganzes stillstand und nur in sich ein chaotisches Durcheinander war. Ich vermochte das ständige Werden der Figuren dann nicht mehr als eine Sukzession in einer bestimmten zeitlichen Richtung aufzufassen, sondern manchmal flossen die Farben und Formen zu einem unbeschreiblichen Durcheinander zusammen, wie wenn die vorher abwechselnden Figuren jetzt alle gleichzeitig erlebt würden*. Hatte ich vorher die Figuren in ständiger Bewegung gesehen, so war jetzt nur noch eine bunte, unsäglich verwirrende Mannigfaltigkeit da, in der ich keine Bewegung mehr wahrnehmen konnte. Wenn ich in die Schau der Figuren ganz versank, so geschah es hier und da, daß ich zugleich auch *in diesen Zeitenstillstand versank, wo das Aufeinanderfolgende in einer stillstehenden Gegenwart zusammenfloß*. Diese Unterbrechungen im normalen Zeiterleben kann ich jetz nicht nur nicht formulieren, sondern mir auch erlebnismäßig kaum mehr vergegenwärtigen. Wenn ich mich von den Figuren losriß und mich gewaltsam der Außenwelt zuwandte, so war dieses abnorme Zeiterlebnis nich mehr vorhanden, aber die Zeitsinnstörung äußerte sich dann in der Täuschung, daß seit dem letzten Aufraffen *ungeheuer lange* Zeit verflossen sein müßte..."

The third illuminating description of time standing still is picked up from ref. 15 (p. 13); this account is interesting by the fact that the author also pays attention to the corresponding distortion in spatial perception (the last line of the vignette):

"Ich merke wohl den Zeitfluß, aber ich erlebte ihn nicht. Ich weiß, daß morgen auch ein Tag ist, ich fühle ihn aber nicht kommen. Die Vergangenheit kann ich jahremäßig abschätzen, ich habe aber keine Beziehung mehr dazu. *Der Zeitstillstand ist unendlich, ich lebe in einer*

* In this and all subsequent excerpts, italics are used to emphasize those parts of the narrative that most directly relate to the subject discussed. English translation of this narrative as well as of other quotations taken from non-english written papers, can be found in the Appendix to the book.

stehenden Ewigkeit. Ich sehe, daß die Uhren sich drehen, aber für mich vergeht die Zeit nicht...

Alles liegt in einer Linie, es sind keine Tiefenunterschiede mehr... Alles ist wie eine feststehende Fläche."

In the last example, the subject makes it clear that it is only their internal time that comes to a complete halt, while the objective/external time still keeps its course. However, there are also experiences where an opposite phenomenon seems to take place, as in the following intriguing case (ref. 132, p. 158):

"Now I was walking and I walked not. And I looked up to the air and saw the air in amazement. And I looked up unto the pole of the heaven and saw it standing still, and the fowls of the heaven without motion. And behold there were sheep being driven and they went not forward but stood still; and the shepherd lifted his hand to smite them with his staff and his hand remained up. And I looked upon the stream of the river and saw the mouths of the kids upon the water and they drank not. And of a sudden all things moved onwards in their course."

4.2 Pure Present of the Second Kind

Time standing still may be viewed as an extreme case of the *de*celeration of the lapse of subjective time, viz. by its infinite slowing-down. This is one limit of the 'present-preserving' disappearance of the flux of time. The other limiting case is represented by an infinite *ac*celeration/speeding-up of time's pace. It is intuitively obvious that this extinction of temporal flow is of a different nature than that characterizing the previous case. And the *second kind* of pure present, Fig. 3h, is indeed found to differ from its 1st kind counterpart. Although it is also a transitory mode, it separates two *identical* modes, Figs. 3a, g; they both are perfect arrows, and of the same polarity at that.

This other kind of pure present is also a well-documented and invariable feature of a certain group of the altered states of consciousness, being most often called an 'eternity', or 'timelessness' [14,18,20,28,36,45–47,52,56–57,64–67,69–70,75–79,83–85,87–94,126,130–131]. It is a sort of telescoping of past, present, and future into the present that is experienced as a 'timeless now', where everything appears to be happening at once. Perhaps the best account of what this is like we found in ref. 76 (p. 46); although the author almost exclusively focuses on the spatial fabric of existence throughout the book, it is the temporal aspect of this particular experience that fascinates most:

"Ich erwachte in einer andern Welt, in der das Welträtsel auf eine unendlich einfache Weise gelöst war in der Art eines andern Raumes. Ich staunte über das Wunder dieses andern Raumes, und in dem Staunen war das Urteil verborgen, er ist grundverschieden von unserem

Raum. Er hatte andere Dimensionen, alles war in allem enthalten. Ich war in diesem Raum, und der Raum war ich. Das Weltall war in dem Raum enthalten, ich im All and es in mir...

Jedenfalls erlebte ich in der zweiten Phase dieses Traumes erst *die Zeit, Weltallzeit, Äonen*. In kosmischen Zeitläufen sah man Welten entstehen, aufblühen wie Blumen, bestehen und wieder vergehen. Es war ein ewiges Spiel, in dem es kein Ende gab. *Blickte man zurück in die Vergangenheit, sah man Äonen, blickte man vorwärts in die Zukunft dehnten sich Äonen in Ewigkeit aus, und die Ewigkeit war im Punk der Gegenwart enthalten.* Man befand sich in einem Sein, in dem das Werden und Vergehen bereits enthalten war, und dieses Sein war mein Bewußtsein. Es enthielt alles..."

Another gripping picture is taken from ref. 130 (Chapt. 2), based on the author's personal near-death experience:

"This time, I moved, not my environment, and I moved rapidly... My speed accelerated until I noticed a wide but thin-edged expanse of bright light ahead, like a "parting" in space or a "lip", with a brightness so brilliant it was beyond light yet I could look upon it without pain or discomfort... The closer I came the larger the parting in space appeared until... I was absorbed by it as if engulfed by a force field...

Further movement on my part ceased because of the shock of what happened next. Before me there loomed two gigantic, impossibly huge masses spinning at great speed, looking for all the world like cyclones. One was inverted over the other, forming an hourglass shape, but where the spouts should have touched there was instead incredible rays of power shooting out in all directions... I stared at the spectacle before me in disbelief...

As I stared, I came to recognize my former Phyllis self in the midupperleft of the top cyclone. Even though only a speck, *I could see my Phyllis clearly, and superimposed over her were all her past lives and all her future lives happening at the same time in the same place as her present life. Everything was happening at once! Around Phyllis was everyone else she had known and around them many others... The same phenomenon was happening to each and all. Past, present, and future were not separated but, instead, interpenetrated like a multiple hologram combined with its own reflection.*

The only physical movement anyone or anything made was to contract and expand. There was no up or down, right or left, forward or backward. There was only in and out, like breathing, like the universe and all creation were breathing - inhale/exhale, contraction/expansion, in/out, off/on."

The third impressive description of a timeless state is taken from ref. 131. Although slightly out of date, the book can still be regarded as a landmark treatise on various aspects of altered states of consciousness. The account in question is from the paper by Pahnke and Richards (ref. 131, p. 403):

"*From the perspective of the Timeless, I could see my life in retrospect and prospect. It was as if it had all been lived through before, as if we had all been here before, and would be here again. There was a strong pre-ordained feeling about this.* I began to see a bit into the future. I understood that I should go back and try to work through some unresolved problems in my relationships with others, and that there would be considerable suffering ahead. My own death was also dimly sensed and strangely accepted. *I saw the unbroken continuity of my*

past with my future, which was not contradicted by the feeling that this present experience would remain with me and bring about deep changes. **The fact that all was pre-ordained did not contradict living in freedom, fighting for truth and against evil…"**

4.3 Past-Present-Past Configuration

Our next discussion is devoted to a past-present-past mode, Fig. 3d. A comparison of the latter figure with Fig. 3a shows that this mode can simply be looked upon as an anomalous arrow lacking the domain of the future, the latter being replaced by the second domain of the past. Moreover, it is also a particular case of the 'fluxless' time; for (see Figs. 3c-e) it represents a borderline between two kinds of the a-w-a configurations, and these have principal arrows of the opposite polarities (compare Figs. 3c and e). A few first-hand accounts given below illustrate sufficiently well this past-prevailing mode. A couple of them are picked up from a very intriguing paper by Franz Fischer[29], which, in our opinion, collects the best descriptions of the schizophrenic's disturbed sense of time. Both the experiences belong to one and the same patient. On one occasion, the patient reports (ref. 29, pp. 560-1):

"Das Denken ist anders, es hat keinen Stil mehr. *Was ist mit der Zukunft? Man kann sie nicht erreichen. Man kann erzählen von der Gegenwart und von der Vergangenheit, vorstellen kann man sich das nicht mehr*. Man kann es sich nicht ausmalen, und man muß zurücksinken. Alles ist wie ein Fragezeichen…

Es zieht mich zurück, ja, wohin? Da, wo es herkommt, dorthin, wo es früher war. *In die Vergangenheit geht es hinein. Das ist so ein Gefühl, als ob man zurückfallen müsse. Das ist das Verschwinderische, das Vergehen. Die Zeit schlupft in die Vergangenheit hinein*, die Mauern sind gefallen. Früher stand Alles fest. Es ist wie zum Greifen, als ob man es wieder herziehen müsse. *Ist das die Zeit? Weit hergerückt!*

Morgens beim Aufwachen, ja wie soll ich sagen, das Verschwinderische ist dann wieder da; da setzt es mir ganz besonders zu! Ob ich weiß, wo ich bin? Ja schon. *Aber das Verschwinderische, daß keine Zeit da ist, und wo man mit der Zeit anpacken soll, und wann gestern war! Da langt es innen in mir aus, immer weiter zurück, - aber wohin? Die Zeit ist im Abbruch.*"

In another situation (ref. 29, p. 563), she puts it as:

"Ich bleibe stehen; *ich bin zurückgeworfen in die Vergangenheit* durch Worte, die man im Saal spricht. Aber das ist doch Alles selbstverständlich, das muß so sein! *Es gibt keine Gegenwart mehr, sondern nur ein Zurückbezogensein, das ist mehr als ein Gefühl, es geht durch und durch…*

Gibt es denn eine Zukunft? Früher hat es für mich eine Zukunft gegeben, jetzt schrumpft sie immer mehr zusammen. Die Vergangenheit ist so aufdringlich; sie wirft sich über mich; sie zieht mich zurück…

Ich bin wie ein brennender Pfeil, den man nach vorn wirft, dann bleibt er stecken, fällt zurück und löscht aus wie in einem luftleeren Raum. Er wird zurückgeworfen. Damit will ich sagen, *daß es keine Zukunft gibt, und daß ich zurückgeworfen werde. Fremde Gedanken kommen herein und treiben mich in die Vergangenheit ab.* Es ist furchtbar. Es geht durch und durch!..."

An almost 'textbook' example has been borrowed from[62]; in this case, a patient suffers an endogenous depression (ref. 62, p. 229):

"Ich kann nichts mehr voraussehen, *als wenn es keine Zukunft mehr gäbe. Früher habe ich mir immer überlegt, was ich nächste Woche tue und was ich mir im Sommer kaufe; jetzt kann ich das nicht mehr, ich meine immer, es hört jetzt alles auf und morgen ist überhaupt nichts mehr.* Ich komme mir vor wie eine Puppe, die leblos ist und der erst wieder neues Leben eingehaucht werden muß. *An die Zukunft kann ich nicht denken - ich kann nicht vorausdenken.*" (Und wie steht es mir der *Vergangenheit?*) "*Ja, an die denke ich sehr stark sogar.* Es ist mir sehr arg, daß die schönen Stunden so vorbei sind." (Und die *Gegenwart?*) "*An die denke ich weniger, die nehme ich halt so hin, der Tag vergeht.* Ich studiere halt immer, wo das alles herkommt." (Wie geht es mit dem Denken?) "Jetzt kann ich schon denken, bloß voraus, das geht einfach nicht, es ist einfach, als wenn es abgeschnitten wäre."

To conclude this subsection, we introduce one more experience where the existence of the two domains of the past is explicitly mentioned (ref. 90, pp. 112-3):

"Every day is a thousands of years, yet the days behind me all collapse into nothing, like a pack of cards. Each day is so long that a normal human being can't imagine it. Each moment is the same way - long. Nothing within this time world has any meaning for me, which is why the time is so long... There are no natural divisions...

I have two pasts, the past before I lost my feelings, which my mind struggles to recapture, and the undifferentiated past of my "time world", which does not interest me. The two pasts are separated by the barrier known as "loss of feelings." The past, or before loss of feelings, is steadily fading in my memory, and since my "new past" is meaningless I don't remember it; so, soon, I will have no past whatsoever, and no future; I will exist only in the endless present..."

4.4 Past-Related Arrow Within Arrow

All the anomalous temporal patterns so far dealt with are characterized by the absence of the flux of time. The remaining two configurations to be discussed, viz. the a-w-a structures, show the contrary. Not only are they characterized by the flux of time, but each of them exhibits a couple of flows - principal and subsidiary, the two pointing against each other (see Figs. 3c, e). Let us examine first the *past*-related mode. As the principal arrow of the latter has the opposite polarity than the reference arrow of our ordinary state (Sec. 3), the psychotic subject experiencing such a configuration will feel that time moves *back*ward. Moreover, as the present of the subsidiary arrow is located in what would otherwise be the domain of the past of the reference arrow, the subject will also report to have a direct access to his *past* events, re-living them in the ordinary sequence (as the subsidiary arrow has the same polarity as the reference arrow, compare Fig. 3c with Fig. 3a). From all the 'psycho-time-related' references[9–133] we have had a possibility to look into, no account draws a portrait of the essential properties of this mode better than that found in ref. 29 (p. 556):

"Gestern am Mittag, als das Essen aufgetragen wurde, schaute ich zur Uhr: Warum sollte man das auch nicht tun, aber da war etwas Besonderes dabei. Denn die Uhr konnte auch nicht mehr helfen und hatte mir nichts mehr zu sagen. Wie sollte ich mich mit der Uhr zusammenbringen? *Ich fühlte mich wie zurückversetzt, als ob Etwas das vergangen sei, gewissermaßen auf mich zukäme, als ob ich selbst einen Weg zurücklegte. Es war mir so, als sei um 11½ Uhr wieder 11 Uhr, aber nicht nur die Zeit war wieder dieselbe und kam zurück, sondern auch, was in dieser Zeit sich für mich zugetragen hatte.* Aber eigentlich ist das Alles viel tiefer, als daß ich es sagen kann. Mitten drin, als ob es nicht dazugehöre, kam Etwas Anderes herauf. *Es war mit einem Male nicht nur wieder 11 Uhr, nein, eine längst vergangene Zeit war da, und darin* - habe ich Ihnen das schon erzählt, vom Samenkorn in der großen, harten Schale? So war das auch wieder: Mitten drin in der Zeit aus die Vergangenheit komme ich auf mich selbst zu. Wie entsetzlich war das. Ich dachte mir, vielleicht ist die Uhr zurückgestellt, da haben die Pfleger einen dummen Streich mit der Uhr gespielt. Ich beschäftigte mich damit, die Zeit als normal zu betrachten, aber es gelang mir nicht; und dann kam ein Gefühl einer furchtbaren Erwartung für mich, *ich könne in das Vergangene hineingezogen werden*, oder das Vergangene könne über mich kommen. Es war unheimlich, daß man so mit der Zeit spielt, irgendwie dämonisch..."

Here, we use a sans serif type style to mark those parts of the narrative that indirectly point out the existence of the subsidiary arrow. A psychotic patient of Laing (ref. 133, p. 148) gives a very succinct description:

"*... ebbi la sensazione che il tempo stesse andando all'indietro; sentii che il tempo procedeva al contrario*, ebbi proprio questa straordinaria sensazione, già... la sensazione più importante in quel momento era quella, *del tempo alla rovescia*... La sensazione era tanto viva che guardai l'orologio e non so come ebbi l'impressione che l'orologio mi confermasse quest'idea, benchè non riuscissi a scorgere il movimento delle lancette..."

A strikingly similar portrayal of time-reversal is also provided by a(nother) depressive patient of Kloos (ref. 62, p. 237):

"Bei meinem plötzlichen Zusammenbrechen war es so in mir, daß die Zeit abgelaufen ist. Ich hatte nach dem Krankenlager nach 3 Wochen das Gefühl, die Uhrzeiger laufen leer, sie haben keinen Halt. Das war mir auf einmal so. Ich fand sozusagen keinen Halt mehr an der Uhr und auch am Leben, ich hatte einen furchtbaren seelischen Zusammenbruch. Wie ich gerade auf die Uhr aufmerksam wurde, weiß ich selber nicht. *Zugleich hatte ich das Gefühl, daß die Uhrzeiger nach rückwärts laufen*... Das bleibt ein einziges Stück sozusagen, und das bleibt stehen. Ich konnte nicht glauben, daß die Zeit wirklich fortschreitet, deswegen meinte ich wohl, die Uhrzeiger hätten keinen Halt und liefen leer... *Wie ich immer arbeitete und arbeitete und mich plagte und nichts zustande brachte, hatte ich eben das Gefühl, es geht mit uns alles zurück... Ich kam einfach in meiner Krankheit nicht mehr mit, und da hatte ich einfach diesen Wahn in mir, daß die Zeit nach rückwärts läuft...* Ich habe mich nicht mehr ausgekannt und hab immer gedacht, ich verlier den Verstand. *Ich glaubte immer, die Uhrzeiger laufen verkehrt, sie sind soviel wie zwecklos. Ich stand gerade auf vom Krankenlager und sah auf die Uhr - da fiels mir auf einmal ein: ja was ist denn das, die Zeit laufen ja verkehrt?!... Ich sah natürlich, daß die Zeiger vorwärts gingen, aber wie ich es nicht glauben konnte, meinte ich immer, die Uhr geht in Wirklichkeit doch zurück...*"

4.5 Future-Related Arrow Within Arrow

To move backward in time is quite a fascinating thing, but going through time into the future is even more astonishing; for the barrier between the present and the future seems to be far more opaque to break than that separating the present from the past. And yet in our last mode, viz. in the *future*-related arrow within arrow (Fig. 3e), this boundary is fully transparent. Really, comparing Fig. 3e with Fig. 3a we see that the present of the subsidiary arrow of Fig. 3e corresponds to an event of the *future* of the reference arrow of Fig. 3a. So, the subject can now glimpse some part of his/her own future; this will be done in a reverse order as, in this case, the polarity of the subsidiary arrow is opposite to that of the reference arrow. In this particular state, there is thus nothing mysterious about such phenomena as clairvoyance (the power of seeing future events in the mind), precognition (the knowledge that something will happen before it does), and/or déjà-vu experiences (the feelings that one has previously experienced exactly the same thing as one is experiencing now). To illustrate the case, it is worth introducing the following premonitory dream of J. W. Dunne (ref. 132, p. 165):

"I seemed to be standing on high ground. Here and there in this were little fissures, and from these jets of vapor were spouting upward. In my dream I recognized the place as an island of which I had dreamed before - an island which was in imminent peril from a volcano. And, when I saw the vapor spouting from the ground, I gasped: 'It's the island! Good Lord, the whole thing is about to blow up!'... All through the dream the *number* of the people in danger obsessed my mind. I repeated it to everyone I met, and, at the moment of waking, I was shouting 'Listen! Four thousand people will be killed unless...'

I am not certain when we received our next batch of papers but when they did come the Daily Telegraph was amongst them, and, on opening the centre sheet, this is what met my eyes: *Volcano Disaster in Martinique. Probable Loss of Over 40,000 Lives.*"

As for the phenomenon of clairvoyance, it can be well pinpointed by the case of a famous Dutch clairvoyant Gerard Croiset (ref. 134, pp. 11-12):

The great gift which the Utrecht paragnost possesses is *being able to break through the time and space barriers.* Telepathy and clairvoyance occur at the present. *But Croiset's precognition powers sweep across time, enabling him to glimpse the future.* Although most of us periodically reveal these abilities, Gerard Croiset regularly sees things not visible to the normal human eye or mind. *His images aren't confined to the limits of conventional time and space. Croiset not only describes events occurring in the present hundreds of miles away,... but constantly dips into the past and future as if into an endless sea.* "*The past, present, and the future are difficult to separate for me*", he says.

Though Professor Tenhaeff has spent nearly half a century studying these phenomena, even he is unsure what causes these impressions. But he is confined that someday soon we shall have the answer. "*It very likely has something to do with a new concept of time*", he explains. " *Today not only our old concept of space requires revising, but even our ideas about time.* "

The future may also be accessible for the subject experiencing the so-called near-death experience (NDE). The first to draw attention to this fact seems to have been Kenneth Ring (ref. 135, p. 183):

"... the material I have collected that bears upon a remarkable and previously scarcely noted precognitive feature of the NDE I have called *the personal flashforward* (PF). *If these experiences are what they purport to be, they not only have extremely profound implications for our understanding of the nature of time but also possibly for the future of our planet...*

Personal flashforwards usually occur within the context of an assessment of one's life during an NDE (i.e. during a life review and preview), although occasionally the PF is experienced as a *subsequent* vision. When it takes place while the individual is undergoing an NDE, it is typically described as an image or vision of the future. *It is as though the individual sees something of the whole trajectory of his life, not just past events*, as some previous accounts have implied. *The understanding I have of these PFs is that to the NDEr they represent events of a conditional future - i.e., if he chooses to return to life, then these events will ensue.* In this sense, from the standpoint of an NDEr, *a PF may be likened to a "memory" of future events...*"

5. FINITE FIELDS AND 'DISCRETE', 'DISORDERED' SUBJECTIVE TIME

The classification of the 'pathological' time configurations given in the preceding section relies on the fact that we change the position of *d*-line in the projective plane whereas the ground field of the latter remains intact, being identical to that of the real numbers. In this section we shall briefly discuss a complementary case. That is, we shall keep *d*-line in a fixed position, and replace the real numbers by some other field(s). In a slightly different context, this case has been dealt with in detail in refs. 4-6, where the interested reader is referred to for the corresponding mathematical background. Here, we shall introduce only those elements of the formalism that are necessary for grasping the idea behind, focussing on the case when *d*-line is in a general position, i.e. when it incorporates neither the base point (B, B′), nor the singular point S.

To this end, it is necessary to generalize our concepts of the past, present and future, given in Sect. 3 for the case of the field of real numbers R, to an arbitrary ground field F. This is quite easy if we take into account factorization properties of the polynomial $P(x)$ represented by eqn. (18). Obviously, if *d*-line is in general position, then both a and b in eqn. (17) are non-zero. As these two parameters are now kept fixed, the factorization properties of the polynomial $P(x)$ depend solely on the value of the parameter ϑ. From what has been said in the first paragraph of Sect. 3 it follows that if this polynomial splits into two distinct factors, two coinciding (repeated) factors, or does not factor at all, then the conic $\mathcal{Q}(\vartheta)$, eqn. (14), is (stands for) a hyperbola (an event of the past), parabola (the present moment), or an ellipse (an event of the future), respectively. And this is the definition of the three temporal domains that can clearly be extended to any field.

Let us now take an event of the R-*future*, for example the one defined by $\vartheta=\vartheta_0=2$ for $\kappa=1$, $a=19$ and $b=3$; then

$$P(x)=x^2+16x+76. \qquad (21)$$

Next, switch to a finite (Galois) field, say to F_7 - the field of residues modulo 7. Since in this field $+16\equiv-5$ and $76\equiv6$, the above polynomial factors as

$$P(x)=x^2+16x+76=x^2-5x+6=(x-2)(x-3) \qquad (22)$$

which implies that the event in question belongs to the F_7-*past*! Moreover, as

$$\tilde{\vartheta}_0=\vartheta+7l, \quad l\in N, \qquad (23)$$

represents one and the same conic in F_7, we see that this past moment of the F_7-temporal corresponds to an *infinite* sequence of events of the R-time: the events with non-negative values of l being in the R-future, while those for which l is negative falling into the R-past. Hence, what we perceive as an ordered sequence of events in our ordinary state, i.e. the state backed by the reals[3], will appear as a disordered, chaotic set in an 'extraordinary' state that corresponds to a modular field[4–6].

This is another fascinating aspect of our theory that finds its counterpart in the realm of psychopathology. Christopher Mayhew, a (former) member of the British parliament, volunteered to participate in a series of laboratory experiments with the drug called mescaline. He reported his experience in the article “An Excursion Out of Time” in the London *Observer* on October 28, 1956, and the way how he perceived time under influence of this drug conforms nicely to the above-outlined pattern (ref. 89, p. 295):

“For half an hour nothing happened. Then I began feeling sick; and various nerves and muscles started twitching unpleasantly. Then, as this wore off, my body became more or less anaesthetized, and I became ‘de-personalized’, i.e., I felt completely detached from my body and the world...

This experience alone would have fully justified the entire experiment for me..., but at about 1.30 all interest in these visual phenomena was abruptly swept aside when I found that time was behaving even more strangely than color. Though perfectly rational and wideawake... *I was not experiencing events in the normal sequence of time. I was experiencing the events of 3.30 before the events of 3.0; the events of 2.0 after the events of 2.45, and so on. Several events I experienced with an equal degree of reality more than once.*

I am not suggesting, of course, that the events of 3.30 *happened* before the events of 3.0, or that any event *happened* more than once. All I am saying is that *I experienced them, not in the familiar sequence of clock time, but in a different, apparently capricious sequence which was outside my control.*

By ‘I’ in this context I mean, of course, my disembodied self, and by ‘experienced’ I mean learned by a special kind of awareness which seemed to comprehend yet be different from seeing, hearing, etc... I count this experience, which occurred when, as I say, I was wide awake and intelligent, sitting in my own armchair at home, as the most astounding and thought-provoking of my life.”

That mescaline intoxication seems to favor this kind of time’s manifold is also evidenced by the following interlude (ref. 77, p. 148):

“... Im Treppensteigen plötzlich wie festgenagelte Momentaufnahme, die momentane Stellung von Dr. M., Dr. St. und mir im Raum aufgefallen. Dies wiederholte sich auf verschiedenen Treppenstufen. Oben angekommen schien *keine Kontinuität der Zeit vorhanden gewesen zu sein, ganzer Vorgang aufgelöst in unzusammenhängende· Einzelsituationen, die nachträglich wie beim Betrachten eines Filmstreifens aktiv daran arbeitend verbunden werden konnten, die aber sowohl im Erleben wie in der unmittelbaren Reproduktion des Geschehens danach durchaus den Charakter des Unabhängigen, Unverbundenen trugen.* Ein seltsames Nebeneinander, nicht Hintereinander, *sie haben keine Stelle in der Zeit,* Zeit hat hier keinen Sinn.”

‘Modular’ time patterns might not also be foreign to other mental psychoses, like, e.g., those suffered by shizophrenics (ref. 78, pp. 104, 106-7):

...Die Unterscheidung von Gegenwart und Zukunft ist ja nicht etwa aufgehoben, da sie doch von beiden Dimensionen immer noch spricht, aber *die Grenzen zwischen real Gegenwärtigem und bloß möglichem, irreal Zukünftigem werden schwankend und überspringbar. Die beiden Dimensionen verschachteln und überlagern sich ohne stetigen Übergang.* Das Zukünftige schlägt ins Gegenwärtige hinein und umgekehrt, und das Erleben bekommt zwischen ihnen etwas flackernd Zwielichtiges, *welches sich in seiner Diskontinuität radikal unterscheidet von der Art, wie etwa der Gesunde in Wachträumen und ähnlichem die Zukunft antizipiert...*

Auch die Grenze von der Gegenwart zur Vergangenheit hin ist schwankend, und dabei wird Vergangenes auf ganz andere Weise in das Gegenwartsgeschehen einbezogen und eingeschmolzen, als auch normalerweise in der Gegenwart das Vergangene mitenthalten ist. *Es ist eine Art von Raffung da; die Gegenwart setzt sich nicht mehr in kontinuierlich-stetigem Fluß von der Vergangenheit ab*, sondern das Gegenwärtige ist zugleich von etwas Vergangenem nicht wie beim Normalen nur durchdrungen, sondern überlagert und überlappt.

In den völlig auf die Zukunft bezogenen... Halluzinationen sahen wir den Schwerpunkt nun eindeutig in das Zukünftige verlagert, welches sinnlich und leibhaftig in die Gegenwart einbrach, *wobei sich die beiden zeitlichen Erlebnisebenen ohne kontinuierlichen Übergang verquickten und verschachtelten.* Durch das Phänomen des zur Realität gewordenen "déjà-vu" erwiesen sich die Grenzen zur Vergangenheit hin ebenfalls als schwankend: *eine der übrigen Vergangenheit nicht integrierte Vergangenheitsdimension drängte sich in diskontinuierlicher Weise ins Gegenwartserleben ein...*

6. DISCUSSION AND CONCLUSION

The findings and results of the previous sections provide us with further evidence, much stronger and more provoking than that already given in refs. 4-8, that *not only* are the manifestations of mental or psychological time so diverse and unusual that they totally fail to conform to the generally adopted picture of the macroscopic physical world, *but* that there also does exist a unique mathematical framework which, at least qualitatively, underlies and unifies their seemingly bizarre properties. Hence, any attempt to disregard these psychopathological temporal constructs as pure hallucinatory phenomena would simultaneously cast a doubtful eye on the very role of mathematics in our understanding of Nature. To the contrary, it is just mathematics (algebraic projective geometry here) that plainly tells us that it is far more natural to expect all these 'unusual' perceptions of time to be simply as real as our ordinary ('normal') one - this being one of the principal messages the paper should convey to the reader. We are, however, fully aware that this point of view is very likely to meet with scepticism, and even with fierce opposition, from the side of 'hard-line' instructional scientists, as envisaged - on a slightly different ground - by St. John (ref. 98, pp. 161-2):

"...the traditional scientific mind refuses to recognize even the possibility of what I am concerned with - the actual nature and quality of the subjective, mystical experience. Materialist scientists of course have no difficulty in recognizing intangible phenomena such as magnetism or cosmic rays passing through a cloud chamber, because they can measure them. Anything that cannot be measured and enumerated doesn't, for the practical purposes of science, exist... By not being able to observe beyond the physical correlates, to reach the internal mental states themselves, science fails to get to grips with what really matters, to recognize that for the individual experiencing them these mental states are much more than a by-product and quite, quite different from a series of (maybe related) biochemical and neurophysiological happenings..."

The main objection will probably be raised against an *anecdotal* character inherent in describing the variety of time's multifaceted phenomena given in the last section. However, we believe this to be the inevitable and the only possible way to follow in pursuit of the research on time in its qualitative aspect, bringing profit to both psychology and physics. As very well argued by Shallis (ref. 132, p. 153):

"Quality and quantity are somewhat like the ingredients of descriptive and instructional science, respectively. Because the two approaches are so different the sorts of evidence employed in each will also differ. Whereas the instructional approach requires, indeed demands, rigorous, quantitative and reproducible evidence, the descriptive attitude, which often deals with the unique and individual, is mainly anecdotal. This does not mean it is uncritical or sloppy, but in trying to find the whole truth everything must be taken into account. If some evidence turns out to be false, that too is part of the picture. In instructional science anecdotal evidence, *even if true*, can be dismissed as unquantifiable and impossible to assess.

The techniques of instructional science cannot handle individual experience or admit to the quality of time. Descriptive science can;..."

We are firmly convinced that anything that shows a definite mathematical structure, whatever bizarre and contraintuitive it may appear, deserves effort and ingenuity to be thoroughly explored and examined, all the more that (ref. 132, pp. 153-4):

"... the fact that the experience of time is not quantifiable puts it into arena of human perceptions that are at once richer and more meaningful than are those things that are merely quantifiable... The lack of quantification of temporal experiences is not something that should stand them in low stead, to be dismissed as nothing more than fleeting perceptions or as merely anecdotal; rather that lack should be seen as their strength. It is because the experience of time is not quantifiable and not subject to numerical comparison that makes it something of quality, something containing the essence of being..."

Last but not least, since the complexity of time is, in our opinion, very intimately tied up with that of consciousness, and the latter seems to be a basic characteristic of the world, the above-discussed geometrical patterns of psycho-(space)time may well turn out to be a decisive clue to breaking the enigma of time; for ref. 136 (p. 1):

"...the creation most suited to probing time's subtle secrets is the human mind, which could, were its cranial computer to be suitably programmed, produce more astounding revelations."

And if nothing else, they should at least stir up some controversy, and perhaps encourage some research along what seems to be an extremely promising line of scientific inquiry.

ACKNOWLEDGEMENTS

I want to thank Mr. Pavol Bendík for painstaking drawing of all the figures, and Dr. Daniel Novocký for (the help with) scanning and conversion of the latter into postscript form. I am also indebted to Dr. Mark Stuckey (Elizabethtown College) for a careful proofreading of the manuscript. Finally, I would like to express my gratitude to Dr. Rosolino Buccheri (CNR, Palermo), Prof. Carl De Vito (The University of Arizona), Prof. Thomas Nordahl (Lawrence Berkeley National Laboratory) as well as to Prof. Jonathan Smith (Iowa State University) for fruitful discussions on various issues pertinent to the topic.

REFERENCES

1. Saniga M., 1996, "Arrow of time and spatial dimensions", *Cosmological Constant and the Evolution of the Universe*, Sato, Suginohara, and Sugiyama (eds.), Universal Academy Press Inc., Tokyo, 283.
2. Saniga M., 1996, "On the transmutation and annihilation of pencil-generated space-time dimensions", *Mathematical Models of Time and Their Application to Physics and Cosmology*, Tifft and Cocke (eds.), Kluwer Academic Publishers, Dordrecht; *Astrophysics and Space Science*, **244**, 283.
3. Saniga M., 1998, "Pencils of conics: a means towards a deeper understanding of the arrow of time?", *Chaos, Solitons & Fractals*, **9** 1071.
4. Saniga M., 1998, "Time arrows over ground fields of an uneven characteristic", *Chaos, Solitons & Fractals*, **9** 1087.
5. Saniga M., 1998, "Temporal dimension over Galois fields of characteristic two", *Chaos, Solitons & Fractals*, **9** 1095.
6. Saniga M., 1998, "On a remarkable relation between future and past over quadratic Galois fields", *Chaos, Solitons & Fractals*, **9** 1769.
7. Saniga M., 1998, "Unveiling the nature of time: altered states of consciousness and pencil-generated space-times", *International Journal of Transdisciplinary Studies*, **2** (2) 8.
8. Saniga M., 1999, "Geometry of psycho(patho)logical space-times: a clue to resolving the enigma of time?", *The Noetic Journal*, **2** (3), 265.
9. Wentzel A., 1922, "Das Weltuntergangserlebnis in der Schizophrenie", *Zeitschr. ges. Neurol. Psychiat.*, **78**.
10. Lehmann H.E., 1967, "Time and psychopathology", *Ann. New York Acad. Sci.*, **138** 798.
11. Binswanger L., 1933, "Das Raumproblem in der Psychopathologie", *Zeitschr. ges. Neurol. Psychiat.*, **145** 598.
12. Cohen J., 1956, "Le temps psychologique", *J. Psychol.*, **53** 285.
13. Guiraud P., 1952, "Les états d'étrangeté", *J. Psychol.*, **45** 449.

14. Gebsattel V.E. von, 1954, *Die Störungen des Werdens und des Zeiterlebens im Rahmen psychiatrischen Erkrankungen*, Springer-Verlag, Berlin-Heidelberg.
15. Tellenbach H., 1956, "Die Räumlichkeit der Melancholischen. I. Mitteilung", *Nervenarzt*, **27** 12.
16. Tellenbach H., 1956, "Die Räumlichkeit der Melancholischen. II. Mitteilung", *Nervenarzt*, **27** 289.
17. Israeli N., 1932, "The psychopathology of time", *Psychol. Rev.*, **39** 486.
18. Schilder P., 1935, "Psychopathologie der Zeit", *Imago*, **21** 261.
19. Häfner H., 1954, "Über Zeitdehnungs- und Zeitbeschleunigungsphänomenen im Rahmen von Zwischenhirnstörungen", *Monatschr. Psychiat. Neurol.*, **127** 336.
20. Stace W.T., 1952, *Time and Eternity*, Princeton UniversityPress, Princeton.
21. Stein J. and Mayer-Gross W., "Über Störungen des Raumsinns und Zeitsinns unter Meskalinwirkung", *Zentralblatt ges. Neurol. Psychiat.*, **45**.
22. Garza C. O de la and Worschel P., 1956, "Time and space orientation in schizophrenia", *J. Abnorm. Soc. Psychol.*, **52** 191.
23. Ehrentheil O.E. and Jenney P.B., 1960, "Does time stand still for some psychotics?", *Arch. Gen. Psychiat.*, **3** 1.
24. Dahl M., 1958, "A singular disturbance of temporal orientation", *Amer. J. Psychiat.*, **115** 146.
25. Dubois F.S., 1954, "The sense of time and its relation to psychiatric illness", *Amer J. Psychiat.*, **111** 46.
26. Straus E.W., 1947, "Disorders of personal time in depressive states", *South. Med. J.*, **40** 154.
27. Weizsäcker V. von, 1924, "Über eine systematische Raumsinnesstörung", *Z. Nervenheilkunde*, **84**.
28. Wallace M. and Rubin A.I., 1960, "Temporal experience", *Psychol. Bull.*, **57** 213.
29. Fischer F., 1929, "Zeitstruktur und Schizophrenie", *Zeitschr. ges. Neurol. Psychiat.*, **121** 544.
30. Fischer F., 1930, "Raum-Zeit-Struktur und Denkstörung in der Schizophrenie", *Zeitschr. ges. Neurol. Psychiat.*, **124** 241.
31. Fischer F., 1933, "Zur Klinik und Psychologie des Raumerlebens", *Schweiz. Arch. Neurol. Psychiat.*, **31** 59.
32. Fischer F., 1933, "Zur Klinik und Psychologie des Raumerlebens", *Schweiz. Arch. Neurol. Psychiat.*, **31** 241.
33. Fischer F., 1934, *"Über die Wandlungen des Raumes im Aufbau der schizophrenen Erlebniswelt"*, Nervenarzt, **7** 84.
34. Scheller H., 1957, "Das Problem des Raumes in der Psychopathologie", *Studium Generale*, **10** 563.
35. Schilder P., 1936, "Psychoanalyse des Raumes", *Imago*, **22** 61.
36. Schilder P., 1936, "Psychopathology of time", *J. Nerv. Ment. Dis.*, **83** 530.
37. Wagner W., 1933, "Über Raumstörung", *Monatschr. Psychiat. Neurol.*, **84** 304.
38. Wolf R., 1952, "Das Raum- und Zeiterleben unter abnormen Bedingungen, besonders im Meskalinrausch", *Deutsch. Med. Wschr.*, **77** 168.
39. Zutt J., 1953, "'Außersichsein' und 'auf sich selbst zurückblicken' als Ausnahmezustand - Zur Psychopathologie des Raumerlebens", *Nervenarzt*, **24**.
40. MacLeod R.B. and Rolf M.F., 1923, "An experiment in temporal disorientation", *Acta Psychol.*, **1** 381.
41. Melges F.T. and Fourgerousse C.E., 1966, "Time sense, emotions, and acute mental illness", *J. Psychiat. Res.*, **4** 127.

42. Hoff H. and Pötzl O., 1934, "Über eine Zeitrafferwirkung bei homonymer linksseitiger Hemianopsie", *Zeitschr. ges. Neurol. Psychiat.*, **151** 599.
43. Ploeger A., 1966, "Zeiterleben in einer Extremsituation", *Zeitschr. Psychotherapie need. Psychol.*, **16** 13.
44. Laubenthal F., 1938, "Zur Pathologie des Raumerlebens unter besonderer Berücksichtigung des Sehraumes", *Zeitschr. ges. Neurol. Psychiat.*, **162** 205.
45. Luner H., 1962, *Die experimentelle Psychose*, Springer-Verlag, Berlin-Göttingen-Heidelberg.
46. Minkowski E., 1933, *Le Temps Vécu - Études Phénoménologiques et Psychopathologiques*, Gauthier, Paris. English translation: 1970, *Lived Time - Phenomenological and Psychopathological Studies*, University Press, Evanston.
47. Minkowski E., 1932, "Le probléme du temps en psychopathologie", *Rech. Phil.*, **2** 231.
48. Morgenthaler W., 1931, *Der Abbau der Raumdarstellungen bei Geistkranken*, Enke, Stuttgart.
49. Kahn E., 1951, "Erwägungen über Raum- und Zeiterleben", *Monatschr. Psychiat. Neurol.*, **122** 1.
50. Klien H., 1919, "Beitrag zur Psychopathologie und Psychologie des Zeitsinns", *Zeitschr. Pathopsychol.*, **3** 307.
51. Mifka P., 1962, "Abnorme Zeiterleben ohne zerebrale Krankheit", *Wien. Zeitschr. Nervenheilk.*, **20** 123.
52. Sturt M., 1925, *The Psychology of Time*, Kegan Paul, London.
53. Reko V., 1938, *Magische Gifte*, Enke, Stuttgart.
54. Lavelle L., 1945, *Du Temps et de l'Éternité*, Aubier, Paris.
55. Pick A., 1919, "Psychopathologie des Zeitsinns", *Zeitschr. Pathopsychol.*, **3** 430.
56. Souriau M., 1937, *Le Temps*, Alcan, Paris.
57. Sivadjan J., 1938, *Le Temps*, Hermann, Paris.
58. Straus E.W., 1928, "Das Zeiterlebnis in der endogenen Depression und in der psychopathischen Verstimmung", *Monatschr. Psychiat. Neurol.*, **68** 640.
59. Borel E., 1949, *L'Espace et le Temps*, Presses Univ. de France, Paris.
60. Cooper L.F. and Erickson M.H., 1954, *Time Distortion in Hypnosis*, Baltimore.
61. Ehrenwald H., 1931, "Störung der Zeitauffassung, der räumlichen Orientierung, des Zeichnens und Rechnens bei einem Hirnverletzten", *Zeitschr. ges. Neurol. Psychiat.*, **132** 518.
62. Kloos G., 1938, "Störungen des Zeiterlebens in der endogenen Depression", *Nervenarzt*, **11** 225.
63. Konczewska H., 1949, "Les métamorphoses de l'espace et du temps dans la mémoire", *J. Psychol. Norm. Pathol.*, **42** 481.
64. Lévy J., 1969, *Le Temps Psychologique*, Dunod, Paris.
65. Schaltenbrandt G., 1963, *Zeit in nervenärztlicher Sicht*, Enke, Stuttgart.
66. Strond J.M., 1960, "The fine structure of psychological time", *Information Theory in Psychology*, Quastler (ed.), Free Press, Glencoe, 174.
67. Coomaraswamy A.K., 1947, *Time and Eternity*, Atribus Asiae Publishers, Ascona.
68. Benda P. and Orsini F., 1959, "À propos de l'étrangeté temporo-spatiale sous LSD", *Ann. Méd. Psychol.*, **117** 525.
69. Bonaparte M., 1939, "L'inconscient et le temps", *Rev. Franc. Psychanal.*, **11** 61.
70. Cohen J., 1967, *Psychological Time in Health and Disease*, Ch. C. Thomas, Springfield.
71. Goody W., 1969, "Disorders of the time sense", *Handbook of Clinical Neurology*, Vinken and Bruyn (eds.), Amsterdam, Vol. 3, 229.
72. Goody W., 1969, "Outside time and inside time", *Persp. Biol. Med.*, **12** 239.
73. Goody W., 1966, "Disorder of orientation in space-time", *Brit. J. Psychiat.*, **112** 661.

74. Köhler W., 1923, "Zur Theorie des Sukzessivvergleichs und der Zeitfehler", *Psychol. Forschung*, **4** 115.
75. Jaspers K., 1923, *Allgemeine Psychopathologie*, Springer-Verlag, Berlin; English Translation: 1968, *General Psychopathology*, The University of Chicago Press, Chicago.
76. Huber G., 1955, *Akasa, der Mystische Raum*, Origo Verlag, Zürich.
77. Beringer K., 1969, *Der Meskalinrausch, Seine Geschichte und Erscheinungsweise*, Springer-Verlag, Berlin-Heidelberg-New York.
78. Ciompi L., 1961, "Über abnormes Zeiterleben bei einer Schizophrenen", *Psychiat. Neurol. Basel*, **142** 100.
79. Grof S., 1976, *Realms of the Human Unconscious. Observations from LSD Research*, Dutton and Comp. Inc., New York.
80. Schmitz H., 1967, *Der Raum: Teil 1 - Leibliche Raum*, H. Bouvier & Comp., Bonn.
81. Schmitz H., 1969, *Der Raum: Teil 2 - Gefühlsraum*, H. Bouvier & Comp., Bonn.
82. Meerloo A.M., 1935, "Über Entwicklung und Störung des Zeitsinns", *Zeitschr. ges. Neurol. Psychiat.*, **153** 231.
83. Fraser J.T. and Lawrence N., 1975, *The Study of Time II*, Springer-Verlag, Berlin-Heidelberg-New York.
84. Hoch P.H. and Zubin J., 1957, *Experimental Psychopathology*, Grune and Stratton, New York.
85. Orme J.E., 1969, *Time, Experience and Behaviour*, Iliffe Books Ltd., London.
86. Hoffer A. and Callbeck M.J., 1960, "Drug-induced schizophrenia", *J. Mental Sci.*, **106**, 138.
87. Arlow J.A., 1986, "Psychoanalysis and time", *J. Amer. Psychoanal. Assoc.*, **34** 507.
88. Melges F.T., 1982, *Time and the Inner Future: A Temporal Approach to Psychiatric Disorders*, John Wiley & Sons, New York.
89. Ebin D., 1961, *The Drug Experience*, The Orion Press, New York.
90. Hartocollis P., 1983, *Time and Timelessness or The Varieties of Temporal Experience*, International Universities Press Inc., New York.
91. Kenna J.C. and Sedman G., 1964, "The subjective experience of time during Lysergic Acid Diethylamide (LSD-25) intoxication", *Psychopharmacologia*, **5** 280.
92. Page J.D., 1969, *Psychopathology*, Aldine Atherton, Chicago.
93. Sims A., 1955, *Symptoms in the Mind. An Introduction to Descriptive Psychopathology*, W. B. Saunders Company Ltd., London.
94. Fraisse P., 1963, *The Psychology of Time*, Harper & Row, New York.
95. Mezey A.G. and Knight E.J., 1965, "Time sense in hypomanic illness", *Arch. Gen. Psychiat.*, **12** 184.
96. Melges F.T. and Freeman A.M., 1977, "Temporal disorganization and inner-outer confusion in acute mental illness", *Amer. J. Psychiat.*, **134** 874.
97. Dilling C.A. and Rabin A.I., 1967, "Temporal experience in depressive states and schizophrenia", *J. Consult. Psychol.*, **31** 604.
98. St. John J., 1977, *Travels in Inner Space*, Victor Gollanz Ltd., London.
99. Koshika K., 1981, "Zur Beziehung zwischen Gehirnpathologie und Psychopathologie im Lichte des Raum-Zeit-Problems", *Bull. Osaka Med. School*, **27** 46.
100. Williams J.M., 1989, "Memory disorder and subjective time estimation", *J. Clin. Exp. Neuropsychol.*, **11** 713.
101. Vroon P.A., 1970, "Divisibility and retention of psychological time", *Acta Psychol.*, **32** 366.
102. Melges F.T., Tinklenberg J.R., Hollister L.E., and Gillepsie H.K., 1970, "Temporal disintegration and depersonalization during Marihuana intoxication", *Arch. Gen. Psychiat.*, **23** 204.

103. Freeman A.M. and Melges F.T., 1978, "Temporal disorganization, depersonalization, and persecutory ideation in acute mental illness", *Am. J. Psychiat.*, **135** 123.
104. Freeman A.M. and Melges F.T., 1977, "Depersonalization and temporal disintegration in acute mental illness", *Am. J. Psychiat.*, **134** 679.
105. Kirstein L. and Buckberg J., 1979, "Temporal disorganization and primary affective disorder", *Am. J. Psychiat.*, **136** 1313.
106. Heim M. and Morgner J., 1985, "Das schizophren gestörte Zeiterleben und seine Beeinflussung durch Neuroleptika", *Psychiat. Neurol. med. Psychol.*, Leipzig, **37** 410.
107. Shoda H., 1993, "Splitting phenomena from a viewpoint of experiencing time: spectrum from multiple personality and hysteria to borderline personality disorder", *Psychopathology*, **26** 240.
108. Mitterauer B., 1983, "Die Logik der wahnhaften Personenverkennung 'Geisteskrankheit' als Störung der subjektiven Zeit", *Zschr. psychosom. Med.*, **29** 85.
109. Sztulman H., 1968, "Psychométrie du temps dans la schizophrénie", *Ann. Méd. Psychol.*, **126** 27.
110. Marchais P., 1977, "De la désorientation temporo-spatiale. I. - L'espace", *Ann. Méd. Psychol.*, **135** 351.
111. Marchais P., 1977, "De la désorientation temporo-spatiale. II. - Le temps", *Ann. Méd. Psychol.*, **135** 898.
112. Tellenbach H., 1990, "Aspekte der Zeit - ihre Deformationen in psychischen Störungen", *Z. Klin. Psychol. Psychopathol. Psychother.*, **38** 52.
113. Mariategui J. and Flores J.R., 1966, "La experiencia del tiempo detenido en la esquizofrenia", *Revista de Neuro-Psiquiatria*, Lima, **29** 301.
114. Mundt Ch., Richter P., Hees van H., and Stumpf T., 1998, "Zeiterleben und Zeitschätzung depressiver Patienten", *Nervenarzt*, **69** 38.
115. Marchais P., 1989, "L'espace-temps en psychiatrie. 1. - Esquisse systémale", *Ann. Méd. Psychol.*, **147** 471.
116. Marchais P., 1989, "L'espace-temps en psychiatrie. 2. - Le champ spatio-temporel", *Ann. Méd. Psychol.*, **147** 570.
117. Marchais P., 1989, "L'espace-temps en psychiatrie. 3. - D'une étude qualitative de la temporalité dans états dépressifs", *Ann. Méd. Psychol.*, **147** 1087.
118. Hellström C. and Carlsson S.G., 1996, "The long-lasting now: Disorganization in subjective time in long-lasting pain", *Scand. J. Psychol.*, **37** 416.
119. Payk T.R., 1977, "Störungen des Zeiterlebens bei den endogenen Psychosen", *Schweiz. Arch. Neurol. Neurochir. Psychiat.*, **121** 277.
120. Giovanardi Rossi P., 1973, "Destructuration de la conscience de l'espace et du temps pendant l'age evolutif", *Revue Neuropsychiat. Infant.*, **21** 63.
121. Michel L., 1981, "Notion de temps chez le schizophrène", *Rev. Méd. Suisse Romande*, **101** 565.
122. Géraud J., Moron P., and Sztulman H., 1967, "A propos d'une étude expérimentale sur l'appréhension du temps et son vécu chez un groupe de schizophrènes", *Ann. Méd. Psychol.*, **125** 802.
123. Muscatello C.F. and Giovanardi Rossi P.,1967, "Perdita della visione mentale e patologia dell'esperienza temporale", *Giorn. Psichiatr. Neuropatol.*, **95** 765.
124. Giovanardi Rossi P. and Frank L., 1969, "Patologia dell'esperienza del tempo e depersonalizzazione nell'adolescenza", *Giorn. Psichiatr. Neuropatol.*, **97** 655.
125. Manfredi M., 1968, "Sulla psicopatologia del tempo", *Osp. Psichiatr.*, **36** 139.
126. Oae H., 1995, "Changes in sequence of time in schizophrenia - from 'severance' to 'collapse' in flux of time", *Psychiatria et Neurologia Japonica*, **97** 405.
127. Guen Le C., 1958, "Le temps figé du schizophrène", *Evol. Psychiatr.*, **28** 701.

128. Loras O., 1969, "Quelques aspects temps-espace-langage en psychopathologie", *Evol. Psychiatr.*, **34** 65.
129. Ajuriaguerra de J., Boehme M., Richard J., Sinclair H., and Tissot R., 1967, "Désintégration des notions de temps dans les démences dégénératives du grand age", *L'encaephale*, **56** 385.
130. Atwater P.M.H., 1988, *Coming Back to Life: The After-Effects of the Near-Death Experience*, Dodd, Mead, and Comp., New York.
131. Pahnke W.N. and Richards W.A., 1969, "Implications of LSD and experimental mysticism", *Altered States of Consciousness*, Tart (ed.), John Wiley & Sons Inc., New York, 399.
132. Shallis M., 1982, *On Time. An Investigation into Scientific Knowledge and Human Experience*, Burnett Books, New York.
133. Laing R.D., 1968, *La Politica dell'Esperienza*, Ed. Feltrinelli, Milano.
134. Pollack J.H., 1964, *Croiset the Clairvoyant. The Story of the Amazing Dutchman*, Doubleday & Comp. Inc., New York.
135. Ring K., 1984, *Heading Towards Omega*, Morrow & Comp., New York.
136. Hope M., 1991, *Time: The Ultimate Energy*, Element Inc., Rockport.

Discrete spacetime: classical causality, prediction, retrodiction and the mathematical arrow of time

GEORGE JAROSZKIEWICZ
School of Mathematical Sciences, University of Nottingham, UK

1. INTRODUCTION

The term *causality* is frequently used in a way which suggests that an intrinsic causal structure underlies the universe. In relativity this is reinforced by the assumption of a metric tensor with a Lorentzian signature. This gives the traditional light cone structure associated with spacelike and timelike intervals, and imposes conditions on the possible trajectories of particles and quantum field theory operator commutation relations.

We shall discuss the idea that *causality is a convenient account designed to satisfy and conform to the patterns of classical logic that the human Theorist wishes to believe underlies the dynamics of space, time and matter.* In this approach causality need not be associated with any *a priori* concept of metric tensor.

This view of causality has been suggested by various philosophers and scientists. Hume argued that causality is a fiction of the mind. He said that causal reasoning is due to the mind's expectation that the future is like the past and because people have always associated an effect with a cause. Kant believed that causality is a category used to classify experience. Lentzen said that causality is a relation within the realm of conceptual objects. Lurchin said that causality is a personal way of thought that is not among our immediate sensual data, but is rather our basic way to organise that data. For Maxwell the principle of causality expresses the general objective of theoretical sciences to achieve deterministic explanations, and according to Heisenberg, causality only applies to mathematical representations of reality and not to individual mechanical systems.

Studies on the Structure of Time: From Physics to Psycho(path)ology
Edited by Buccheri *et al.*, Kluwer Academic / Plenum Publishers, New York, 2000.

2. THE PPM VIEW OF TIME

To avoid confusion we distinguish three sorts of time:

1. *Process Time* is the hypothesised time of physical reality. Although there is geological and astrophysical evidence for some sort of temporal ordering in reality[1], process time need not exist in any real sense and may just be a convenient way for humans to think about the Universe.
2. *Physiotime* is the subjective time which humans sense and which they believe runs monotonically forwards for them. It is the end product of complex bio-dynamical processes occurring in process time. Its origins are not understood currently. Many physicists believe that this feeling is an illusion. What matters here is the undeniable *existence* of this feeling, because humans are driven by this sensation of an ever increasing time to believe that descriptions of reality must involve such a concept.
3. *Mathematical times* are conceptual inventions of human theorists designed to model process time. Examples are Newtonian Absolute Time, relativistic coordinate times, proper time and cosmic time. Mathematical times usually involve subsets of the real line, which has an ordering property. This ordering is used to model the notions of *earlier* and *later*. This presupposes something about the nature of process time which may be unwarranted. In the Euclidean formulation of field theory for example there is no dynamical ordering parameter.

2.1 The Theorist

We shall use the term *Theorist* to denote the human mind operating at its clearest and most rational in physiotime. The Theorist has the status of an observer or deity overseeing the mathematically consistent development of chosen mathematical models that are used to represent phenomena in process time. *Free will* enters into the discussion here as the freedom of the Theorist to choose boundary conditions in these models. Whether free will is an illusion or not is regarded here as irrelevant.

3. CLASSICAL CAUSALITY

The need to seek causal explanations stems from the peculiarities of human consciousness. Humans generally want to *explain* phenomena. When they do this they invariably try to invoke what may be called *classical logic*. This is the everyday logic that postulates that statements are *either true or else false* and that conclusions can be drawn from given premises. It is also the logic of vision, which generally informs the brain that an object either is in a place or else is not in that place. The rational conscious mind tends to believe that the external universe follows this logic, and this is the basis for the construction of CM (classical mechanics) and all the belief structures which it encodes into its view of reality. It is also the logic of jurisprudence and common sense. This logic served humanity extremely well for millennia, until technological advances in the early years of the twentieth century revealed that quantum phenomena did not obey this logic in detail.

A CM Theorist is anyone who believes in a classical view of reality. In the mindset of a CM Theorist, reality is assumed to be strictly single valued at each and every time even in the absence of observation. Philosophers say that reality is determinate. The CM Theorist attempts to make unique predictions wherever possible, such as where a planet will be at a future time. The assumption is made that the planet will be somewhere at that time and not nowhere, and that it will not be in two or more places at that time.

In general, quantum theory requires a pre-existing classical conceptual framework for a sensible interpretation. For example, relativistic quantum field theory assumes a classical Lorentzian metric over spacetime, and only the fields are quantised. For this reason, we shall focus our attention on a classical formulation of causality.

4. FUNCTIONS AND LINKS

To set up our framework for causality, it will be useful to review the definition of a *function*:

Definition 1: A *function* f is an ordered triple[2] $f \equiv (F, D, R)$ where F, D and R are sets which satisfy the following:

1. F is a subset of the Cartesian product $D \times R$;
2. for each element x in D there is exactly one element y in R such that the ordered pair (x, y) is an element of F.

Physicists tend to write $y=f(x)$. D is called the *domain* of (definition of) f and R is the *range* of f. The *image* of f is the subset $f(D)$ of R such that for each element v in $f(D)$ there is at least one element u in D such that $v=f(u)$.

Without further information it cannot be assumed that $f(D)=R$; but in our work this must be assumed to hold. Otherwise, there arises the possibility of having a CM where something could happen without a cause, i.e. an element z of R could exist for which there is no x in D such that $z=f(x)$.

The range and domain of a function do not have a symmetrical relationship and the ordering of the sets in *Definition* 1 is crucial. Usually, no pair of component sets in the definition of a function can be interchanged without changing the function. This asymmetry forms the basis of the time concept discussed in this article and defines what we call a *mathematical arrow of time*. Assuming that f is single valued, then we may employ the language of dynamics here, though this may seem unusual. We could say that $y=f(x)$ is *determined* by x *via the process* f, or that x *causes* $f(x)$. Then x is a *cause*, $f(x)$ is its *effect* and f is the *mechanism of causation*.

Although *Definition* 1 carefully excludes the concept of a many-valued explicit function, such a possibility arises when we discuss implicit functions.

Graphical notation: We shall use the following graphical notation:

4.1 Explicit functions

The process of mapping elements of D into R via f will be represented by the LHS (left hand side) of Fig.1a, where the large circles denote domain and image sets, the small circle denotes the function, and arrows indicate the direction of the mapping. An alternative representation is given by the RHS (right hand side) of Fig.1a, where the labels D, R and f are understood. Here the integers 0 and 1 represent the ordering of the function *from* D ($\equiv 0$) *to* R ($\equiv 1$), so that the arrows are not needed.

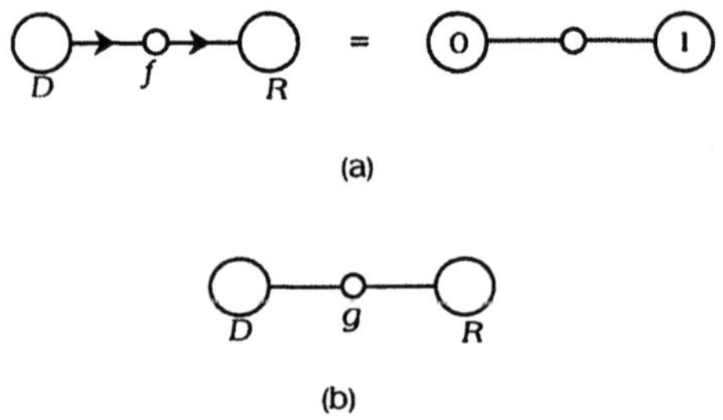

Figure 1. (a) The LHS represents a function f with domain D and range R; the RHS is an equivalent representation of the same with D, R and f understood. The ordering of the digits 0, 1 represents the direction of the mapping. (b) A representation of an implicit function g of two variables.

4.2 Many-to-one functions

Definition 1 raises the question: *which particular* x *in* D *caused a given* y *in* R? As given, nothing in *Definition* 1 rules out many-to-one mappings, so without further information about the function f, there may be more than one such x. The assumption of a unique pre-image is equivalent to the belief that, given the present state of the universe, there was a unique past which gave rise to it. 't Hooft has recently discussed this in the context of gravitation using equivalence classes of causes[3]. This is bound up with the notion of irreversibility.

If however it is the case that f is one-to-one and onto, then its inverse function f^{-1} exists and so any element $y=f(x)$ in R can be mapped back to a unique cause x in D via f^{-1}. In such a case there would be no inherent difference in principle between the roles of D and R. In the language of dynamics the mechanics would be *reversible*.

4.3 Implicit functions and links

Suppose now that the relationship between D and R is implicit rather than explicit. For example, let D and R each be a copy of the real line R with elements x, y belonging to D and R respectively and suppose that the only dynamical information given was an implicit equation of the form

$$g(x, y)=0 \tag{1}$$

Then our graphical notation for this is given by Fig.1b, where the small circle now denotes an implicit equation or *link* g relating elements of D and R. Without further information no arrows are permitted at this stage.

4.4 Resolution of links

Given an implicit equation in two variables such as (1), suppose now that it could proved that there was always a unique solution for y in R given any x in D (the solution y of course depending on the value of x). Then our convention is that now arrows pointing from D into R via the link g may be added to indicate this possibility, giving the LHS of Fig.1a with f replaced by g. But now the relationship between D and R is formally equivalent in principle to having an explicit function y of x (even if y could not be obtained analytically). In such a case we will also say that any given element of D *causes* or *determines* a corresponding element of R and that the link between D and R may be *resolved* from D *to* (or in favour of) R.

Our concept of resolution depends only on the *existence* of a unique solution and does not imply that a solution could actually be computed by the Theorist in practice. Computability is an attribute associated with *physiotime*, and is not here regarded as an essential ingredient of our version of causality. A more severe definition of causality however might be to impose the restriction of computability. We do not do this here because we wish to avoid anthropomorphism. What is important in classical mechanics is the existence of a unique resolution; the universe does not actually "compute" anything when this resolution occurs.

4.5 Inequalities

It is possible to consider links which are not equations but more general relations such as inequalities. For example, suppose D and R are copies of the real line with elements denoted by x, y respectively and consider a link defined by

$$g(x, y) \equiv x+y<0. \tag{2}$$

Given x there is an infinity of solutions for y, so in this case one possible interpretation would be that the link is equivalent to a many valued function of x, although this way of putting things might seem unusual. On the other hand, given a y there is also an infinity of solutions for x. This leads to an alternative interpretation of the link as a many valued function of y.

Such examples do not generate a classical TRA (temporal resolution of alternatives) and so would not occur normally in our classical spacetime dynamics.

4.6 Reversibility

Now suppose we were given an implicit equation for x and y such that we could prove the existence of a unique solution for either x or y given the other variable. Then the arrows could point either way and this would correspond to a choice of causation*. Because cause and effect can be interchanged in such a case, it then becomes meaningful to talk about this dynamics being *reversible*. Clearly this is formally equivalent to having an invertible explicit function.

When discussed in this way it becomes clear that in general, spacetime dynamics will be irreversible. Reversibility will occur only under very special conditions, which of course is the experience of experimentalists.

4.7 Generalization

In general the sets D and R need not be restricted to the reals. They could be any sort of sets, such as vector spaces, tensor product spaces, quaternions, operator rings, group manifolds, etc. Whatever their nature, they will always

* This choice is taken by the Theorist in physiotime.

be referred to as *events* for convenience. In the sorts of dynamics we have in mind events will often be sets such as group manifolds.

Links are defined as specific relationships between events. They may be more complicated than those in the above examples, and may have several or many components relating different components of different events. The specification of a set of events and corresponding links will be said to specify a (classical) *spacetime dynamics*.

An important generalisation is that a link might involve more than two events. Given for example five events R, S, T, U and V with elements r, s, t, u and v respectively then a classical dynamics involving these events would give some set of equations or link g of the generic form

$$g(r, s, t, u, v)=0. \tag{3}$$

This will be represented by Fig.2a.

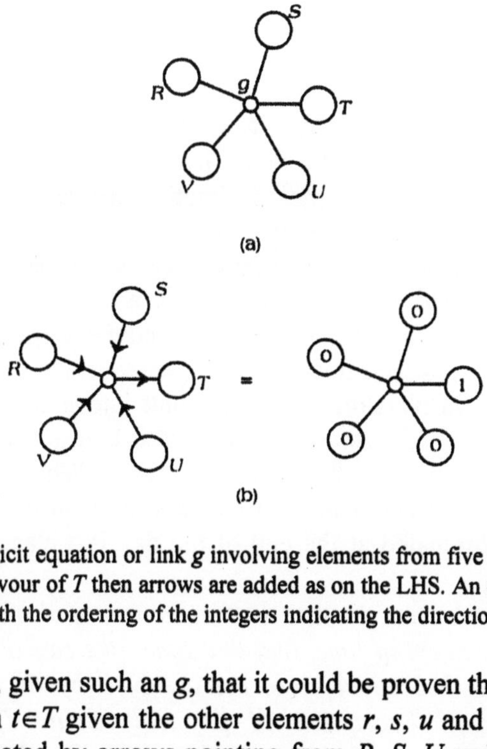

Figure2. (a) An implicit equation or link g involving elements from five events. (b) If the link can be resolved in favour of T then arrows are added as on the LHS. An equivalent diagram is given on the RHS with the ordering of the integers indicating the direction of the resolution.

Suppose now, given such an g, that it could be proven that there is always a unique solution $t \in T$ given the other elements r, s, u and v. In such a case this will be indicated by arrows pointing from R, S, U and V into g and an arrow pointing into T as in the LHS of Fig.2b. Then it will be said that g *can be (causally) resolved in (favour of) T*, and T will be called the *resolved event*.

By definition, classical resolution is always in favour of a single resolved event, given initial data about the other events associated with the link. This does not imply anything about the possibility of resolving g in favour of any of these other events. It may be or not be possible to do this.

Suppose the Theorist could in principle resolve T if they were given R, S, U and V, and also resolve S if they were given R, T, U and V. Then the Theorist has to make a choice of resolution and choose one possible resolution and exclude the other. It would not be meaningful classically to resolve g in favour of T and S at the same moment of physiotime, the reason being that these alternative resolutions employ different and inconsistent initial data sets (boundary conditions). Initial data sets are equivalent to information, and it is a self-evident premise that a Theorist can have at most one initial data set from a collection of possible and mutually inconsistent initial data sets at a given moment of physiotime.

There remains one additional exotic possibility. If might be the case that given say R, S and T, both U and V were implied by a knowledge of the link. For example, suppose the link was equivalent to the equation

$$r+s+it+u+iv=0, \tag{4}$$

where $i=\sqrt{-1}$. Assuming that r, s, t, u and v were always required to be real, then we could always find u and v from a given r, s and t, simply by equating real and imaginary parts. A situation where a given link can be resolved in two or more sets given just one initial data set at that link will be called a *fluctuation process*. Fluctuation processes will be excluded from our notion of classical causality. They may have a role in the QM (quantum mechanics) version of causality, which is beyond the scope of this article.

One reason for excluding fluctuation processes is that this guarantees that information flows (in physiotime) from a link into a *single* event. This is related to the concept of *cosmic time* discussed below and to the idea that classical mathematical time has one dimension. When theorists discuss models with more than one parameter called a time, all but one of these has to be hidden or eliminated at the end of the day if a classical picture is to emerge.

It may be argued therefore that *the mechanism of classical resolution is the origin of the concept of time, and that time, like causality, is a no more than a convenient theoretical construct designed by the human mind to provide a coherent description of physical reality*. This carries no implication that what we called process time is really a linear time. As we said before, process time is just a convenient label for something which may be quite different to what we believe it to be.

When a choice of resolution exists and is made, then as an additional simplification and provided there is no confusion with other relations to which *T* may be a party (not shown), the diagram on the LHS of Fig.2b may be replaced by the RHS of Fig.2b. Here the numbers zero and one indicate the ordering of the resolution. Because *T* may be regarded as *caused* by the other events, it can be regarded as later and so has a greater associated discrete time. Such times will be called *dates*.

In general a link may be a whole collection of relations and equations. If there is just one small part of these equations which does not determine a unique solution fully, i.e., prevents a resolution, then arrows or dates are in principle not permitted. However, under some circumstances it may be reasonable to ignore some part of a dynamical relation in such a way that arrows could be justified as far as the remaining parts were concerned. For example, the microscopic laws of mechanics appear to be resolvable forwards and backwards in time (i.e., are reversible) provided the "small" matter of neutral kaon decays and the thermodynamic arrow of time (which could involve the gravitational field[4] are ignored.

In Fig.1a, *D* is the *complete cause* of *R*; in Fig.2b, *R* is a *partial cause* of *T*. The *complete cause* of *T* is the collection of events *R*, *S*, *U* and *V*, but only for this choice of resolution. The Theorist could decide to alter boundary conditions so that *T* was no longer regarded as the resolved event.

Having outlined our ideas on functions and links, we shall apply them now to discrete spacetime.

5. DISCRETE SPACETIME

Fourier's *principle of similitude* states that a system *S'* similar to but smaller than another system *S* should behave like *S*. It is the physicist's analogue of continuity in mathematics, and is of course an erroneous principle when applied to matter, as evidenced by the observation of atoms and molecules. It is generally supposed that this principle will also break down in the microscopic description of space and time. Classical GR (general relativity) may therefore be an approximation, albeit a remarkably good approximation, to some model of space and time which is not intrinsically a four-dimensional pseudo-Riemannian manifold.

There have been numerous suggestions concerning the fundamental nature and meaning of space and time, such as twistor theory, point set theory, etc.,[5] and each of these suggestions makes a specific set of mathematical assumptions about spacetime. Likewise, in this article a specific view of space time and dynamics is proposed and its consequences explored. Of course, there is an important question concerning the use of classical or quantum physics here. In this paper the proposals are based on classical ideas and the ramifications of quantum physics are explored elsewhere.

From before the time of Newton, physicists took the view that material objects have definite spatial positions at definite times. In the 20th Century theorists went further and developed to the extreme the Wellsian or block Universe[6] view that space and time exist in some physically meaningful sense, even in the absence of matter. Whatever that sense is, be it a physical one or simply an approximate relationship of sorts between more complex attributes of reality, most physicists agree on the prime status of spacetime as the arena in which or over which physical objects exist. This is certainly the case for classical physicists and to various degrees for quantum physicists.

In this paper the focus is on a classical description of a *discrete* spacetime structure. Discreteness is considered here for several reasons. First, as mentioned above, it would be too much to hope that Fourier's principle of similitude should apply to spacetime and not to matter. Second, there is a strong feeling in the subject of quantum gravity that the Planck scales are significant. Third, discreteness has the advantage over continuity in being less mathematically restrictive. Theories based on discrete principles can usually encompass the properties of those based on continuous principles via appropriate limit processes, yet retain features which cannot occur in the continuum.

Discrete spacetime structure and its relationship to causality has been discussed by a number of authors, notably by Sorkin et al.[7] and Finkelstein et al.[8]. A basic difference between these approaches and that taken here is that no *a priori* underlying spacetime manifold is assumed.

Proposition 1: classical spacetime may be modelled by some discrete set S.

Elements of S will be denoted by capital letters such as P, Q, R, etc. S will be called a *spacetime* and its elements referred to as *events* even if subsequently S turns out to have only an indirect relationship with the usual four-dimensional spacetime of physics. What these events mean physically depends on the model. It is simplest to think of events as labels for mathematical structures representing the deeper physical reality associated with process time.

Events are meaningful only in relationship with each other and it is meaningless to talk about a single event in spacetime without a discussion of how the Theorist relates it to other events in spacetime. This is done by specifying the *links* or relationships between the events. Links are as important as the events themselves and it is the totality of links and events which makes up our spacetime dynamics. This should include all attributes relating to matter and gravitation, and it is in principle not possible to discuss one without the other.

The structure of our spacetime dynamics is really all there is; links and events. No preordained notion of metrical causality involving spacelike and timelike intervals is assumed from the outset. All of that should emerge as part of the implications of the theory. In classical continuous spacetime theories, on the other hand, the metric is usually assumed to exist independently of any matter, even before it is found via the equations of GR. As we said before, this metric carries with it lightcone structure and other pre-ordained attributes of causality.

Discrete spacetime carries with it the astonishing possibility of providing a natural explanation for length, area and volume. According to this idea, attributed to Riemann[7], these are simply *numerical counts* of how many events lie in certain subsets of spacetime. Discreteness may also provide a natural scale for the elimination of the divergences of field theory, and permits all sorts of novelties to occur which are difficult if not impossible to build into a manifold. Recently, the study of spin networks in quantum gravity has revealed that quantization of length, area and volume can occur[9].

6. EVENT STATE SPACE

In CM the aim is usually to describe the temporal evolution of chosen dynamical degrees of freedom. These take on many possible forms, such as position coordinates or various fields variables such as scalar, vector and spinor fields. In our approach, we associate with each event P in a given spacetime S an internal space Ω_P of dynamical degrees of freedom called *event state space*. This space could be whatever the Theorist requires to model the situation. Moreover, it could be different in nature at each event P in this spacetime. Elements of Ω_P will be denoted by lower case Greek letters, such as ξ_p, λ_P etc. and a chosen element ξ_P of Ω_P will be called a *state of the event P*.

Elementary examples of event state spaces are:

1. **Scalar fields:** a real valued scalar field f on a spacetime S is simply a rule which assigns at each event $P \in S$ some real number f_p. This may be readily generalised to complex valued functions. If no other structure is involved then obviously

$$\Omega_P = \mathsf{R}_P \ \forall P \in S,$$
$$\xi_P \equiv f_p \in \mathsf{R}_P, \qquad (5)$$

where R_P is a copy at P of the real line R.
A classical configuration of spacetime in this model would then be some set $\{f_P, f_Q, ...\}$. This configuration "exists" at some moment of the Theorist's physiotime but the model itself would not necessarily have any causal ordering. That would have to be determined by the Theorist in the manner discussed below.

2. **Vector fields:** suppose at each event P there is a copy V_P of some finite dimensional vector space V with elements $\mathbf{v} \in V$, etc. Then a vector field is simply a rule which assigns at each event P some element $\mathbf{v}_P \in V_P$.
3. **Group manifolds:** at each event P we choose a copy G_P of some chosen abstract group, such as Z_2 or $SU(3)$ to be our event space.
4. **Spin networks:** A spin network is a graph with edges labelled by representations of a Lie group and vertices labelled by intertwining operators. Spin networks were originally invented by Penrose in an attempt to formulate spacetime physics in terms of combinatorial techniques[10] but they may also defined as graphs embedded in a pre-existing manifold[11]. The state event space at each event P is a copy of $SU(2)$. In this particular model the sort of events we are thinking of in spacetime would be associated with the geometrical links of a triangulation and the links (the dynamical relationship between our events) would be associated with the geometrical vertices of the triangulation, that is, with the intertwining operators.

6.1 Neighbourhoods and local environments

For physically realistic models the number of events in the corresponding spacetime S will be vast, possibly infinite. Sorkin et al.[7] give a figure of the order 10^{139} per cubic centimetre-second, assuming Planck scales for the discrete spacetime structure. We shall find it useful to discuss some examples with a finite number of events for illustrative purposes.

In our spacetime diagrams we will follow the convention established for functions in the previous section; large circles denote events and small circles denote links, with lines connecting events and links. Before any temporal resolution is attempted, no arrows can be drawn. Fig.3a shows a finite spacetime with 14 events and 9 links.

Definition 2: The *(local) environment* $\mathfrak{E}_A$ *of an event A* is the subset of links which involves A, that is, all those links to which A is party, and the *degree of an event* is the number of elements in its local environment.

For example, from Fig.3a, the local environment of the event labelled P is the set of links $\mathfrak{E}_P \equiv \{f, g, h\}$, and so P is a third degree event.

Definition 3: The *neighbourhood* N_A *of an event A* is the set of events linked to A via its environment.

For example, from Fig.3a the neighbourhood of event P is the set of events $\mathcal{N}_P \equiv \{Q, R, S, T, U, V\}$, and Q, R, S, T, U and V are the *neighbours* of P.

Definition 4: The *domain* $\mathcal{D}_f$ *of a link* f is the set of events involved in that link, and the *order of a link* is the number of elements in its domain.

For example, from Fig.3a, $\mathcal{D}_f \equiv \{Q, R, P, U\}$ and so f is a fourth order link.

The local environment of an event will be determined by the underlying dynamics of the spacetime, i.e. the assumed fundamental laws of physics. Currently these laws are still being formulated and discussed, so only a more general (and hence vague) discussion can be given here with some simple examples. A spacetime and its associated structure of neighbourhoods and local environments will be called a *spacetime dynamics*.

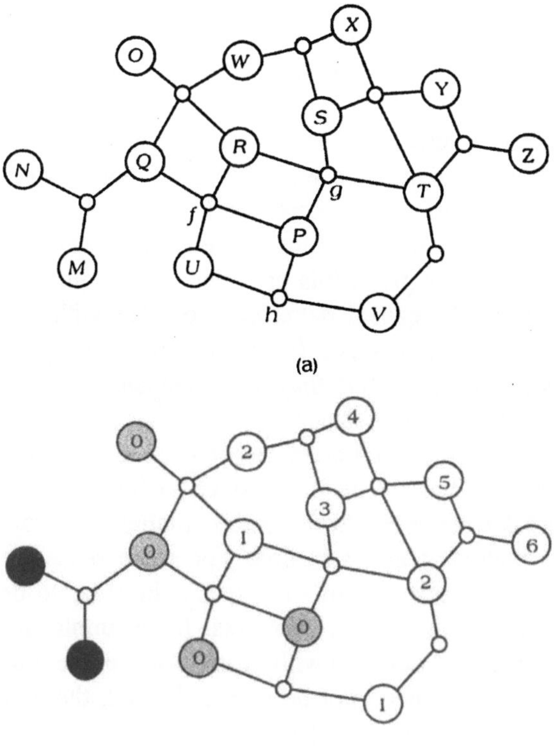

Figure3. (a) A discrete spacetime with 14 events and 9 links. (b) The chosen initial data set is shaded grey and its full implication (or future) is dated.

7. KINDS OF SPACETIME DYNAMICS

There are two interrelated aspects of any spacetime dynamics: i) the nature of the event state space associated with each event and ii) the nature of the links connecting these events. The way in which events are related to each other structurally via the links will be called the *local discrete topology*. If this discrete topology has a regularity holding for all links and events, such as that of some regular lattice network, then we shall call this a *homogeneous* discrete topology. Otherwise it will be called *inhomogeneous*.

We envisage three classes of classical spacetime dynamics:

1. **Type A:** spacetime dynamics with a homogeneous and fixed discrete topology with a variable* event state configuration which does not affect the discrete topology. This corresponds to (say) field theory over Minkowski spacetime;
2. **Type B:** spacetime dynamics with inhomogeneous but fixed discrete topology with a variable event state configuration which does not affect the discrete topology. This corresponds to field theory over a fixed curved background spacetime, such as in the vicinity of a black hole;
3. **Type C:** spacetime dynamics with discrete topology determined by the event state configuration; this corresponds to GR with matter.

Type *A* and *B* spacetime dynamics are relatively easy to discuss. Once a fixed discrete topology is given this provides the template or matrix for the Theorist to "slot in" the causal patterns associated with initial event state configurations (initial data sets). In this sense, types *A* and *B* are not genuinely background free, but they are independent in a sense of any preordained Lorentzian metric structure,

Type *C* presents an altogether more interesting scenario to discuss and demonstrates the basic issue in classical GR which is that the spacetime dynamics should determine its own structure, including its topology.

GR without matter of any sort does not make sense in our approach, because we need to specify event state spaces in order to define the links. Spins are needed to specify spin networks, for example. In our approach, gravitation is intimately bound up with discrete spacetime topology.

A spacetime diagram need not be planar. Indeed, there need not be any concept of spacetime dimension at this stage. Sorkin et al.[7] suggest that at different scales, a given discrete spacetime structure might appear to be approximated by different continuous spacetime dimensions, such as 26, 10, or 4, depending on the scale.

* i.e., variable in physiotime.

8. CLASSICAL RESOLUTION AND CAUSAL STRUCTURE

The PPM model was introduced as an approach to the modelling of process time. Working in physiotime, the Theorist first decides on a spacetime dynamics and from an initial data set then determines a consistent event state at each event in the spacetime. Because of the existence of the links, however, these event states cannot all be independent, and this interdependence induces our notion of causality, as we now explain.

First, restrict the discussion to Type A and B spacetimes and suppose that each link is *fully resolvable*. By this is meant that if the order of a link is n and the event states are specified for any $n-1$ of the events in the domain of the link, then the remaining event space in the domain is uniquely resolved.

Example 1: An example of a fully resolvable link is the following: Let f be a link of order r, with domain $\mathcal{D}_f \equiv \{P_1, P_2, \ldots, P_r\}$. Suppose the event space at P_i is a copy G_i of some group G such as Z_2 or $SU(n)$ and suppose the link is defined by

$$f{:}g_1g_2\ldots g_r=e, \tag{6}$$

where $g_i \in G_i$ and e is the group identity*. Then clearly,

$$g_1=g_r^{-1}g_{r-1}^{-1}\ldots g_2^{-1} \tag{7}$$

and similarly for any of the other events states. In other words, we can always resolve any one of the events in $\mathcal{D}_f$ uniquely in terms of the others.

Now suppose the Theorist chooses one link, such as f in Fig.3a, which is a fourth order link. Then its domain $\mathcal{D}_f$ can be identified immediately from the spacetime dynamics to be $\{P, Q, R, U\}$. The Theorist is free to specify the states at three of these without any constraints, and this represents an initial data set called S_0. Suppose these are events P, Q and U.

If now the structure of the link f is such that there is only one possible solution in $\Omega(R)$, then we have a classical resolution of alternatives† and the emergence of a causal structure. We can use the language of dynamics and say that events P, Q and U *cause* R and denote it by a diagram such as Fig.2b. But it should be kept in mind that the Theorist has *decided on which three sets to use as an initial data set*. Our interpretation of causality is that it is dictated partly by the spacetime dynamics and partly by the choices made

* We may use a matrix representation to define products of elements from different copies of G.

† i.e., the alternatives which form the event state space $\Omega(R)$.

by the Theorist. In general, classical resolution involves using information about $n-1$ event states at an nth order link to determine the state of the remaining event in the domain of that link.

An initial data set may involve more than one link, such as shown in Fig.3b. In that diagram, events are labelled by integers representing the discrete times at which their states may be fixed. Events in the chosen initial data set are labelled with a time 0 and shaded grey. These are the events $\{O, P, Q, U\}$.

Given an initial data set S_0, the Theorist can then use the links to deduce the *first (or primary) implication* S_1. This is the set labelled by integer time 1 in Fig.3b, and consists of events $\{R, V\}$. The event state at each of the events in the primary implication is determined from a knowledge of the event states on the initial data set, assuming the links do indeed permit a classical TRA. It could be the case that the primary implication is the empty set.

Given a knowledge of the event states on S_0 and its primary implication S_1, then the *second (or secondary) implication* $S_2 \in \{W, T\}$ can now be found and its events labelled by the discrete time 2. This process is then repeated until the *full implication* S_∞ of S_0 is determined. In Fig.3b this is the set $\{R, S, T, V, W, X, Y, Z\}$.

In this example it is assumed that each of the links is fully resolvable. If any link is not fully resolvable, it may still be possible to construct a non-empty full implication for certain initial data sets.

Several important concepts can be discussed with this example.

8.1 The future of an initial data set

The full implication of an initial data set consists of those events whose event states follow *from* given initial conditions, and so it is not unreasonable to call the full implication of any initial data set the *future* of that set. Whilst the term "*from*" in the preceding sentence refers to a process of inference or implication carried out in physiotime by the Theorist, the result is that diagram Fig.3b for example now carries *dates* (or equivalently *arrows*) as a consequence. This ordering may now be regarded as an attribute of the mathematical model rather than of physiotime, and this is the *mathematical arrow of time* referred to previously. The resulting structure can then be used to represent phenomena in process time.

However, some caution should be taken here. We may encounter spacetime dynamics which are equivalent to *reversible dynamics* in continuous time mechanics. The full implication of some initial data sets for such spacetime mechanics may propagate both into what we would normally think of as the conventional future *and* into the conventional past. This is in accordance with the general problem encountered with any reversible dynamics: we can never be sure which direction is the real future and which is the real past unless external information is supplied to tell us.

An example of a reversible discrete topology is given in Fig.4a. This is the topology of the discrete time harmonic oscillator[12]. Assuming the links are fully resolvable, we see that the initial data set shown (shaded and labelled 0) has a full implication which extends to the right and to the left of the diagram Fig.4b, i.e. into the conventional past and future.

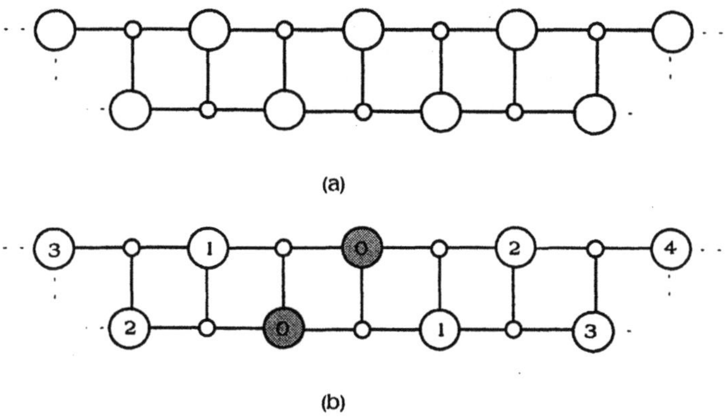

Figure 4. An example of a reversible spacetime dynamics, the discrete time harmonic oscillator. (a) The basic discrete topology. (b) The chosen initial data set (shaded) needs two real numbers, equivalent to initial position and velocity. The full implication runs into the left (conventional past) and to the right (conventional future), and there are no inaccesible events.

8.2 Inaccessible events

In Fig.3b the set $\{M, N\}$ is inaccessible from the given initial data set $\{O, P, Q, U\}$, that is, its intersection with both the initial data set and its full implication is the empty set. Such an inaccessible set may be interpreted in a number of ways. It could be thought of as the *absolute past* of the initial data set $\{O, P, Q, U\}$ because in this particular example, the initial data set implies nothing about $\{M, N\}$, but specifying the states at M and N would fix Q. In other examples such inaccessible events could be interpreted as beyond an event horizon of some sort. The general feature of inaccessible events is that they cannot be affected by any changes to the event states in an initial data set.

Archaeology is our term for the process of reconstructing portions of an absolute past from new and limited information added to some initial data set. It means the process whereby specifying one or more event states in an absolute past has the immediate consequence that the Theorist can deduce even more information about that past.

Example 2: Consider a spacetime dynamics based on the infinite regular triangular topology shown in Fig.5, with an initial data set shaded and labelled by 0. Its first, second and third implications are labelled by 1, 2 and 3 respectively. The full implication of the initial data set in fact extends to infinity on the right. Suppose the Theorist now determines in some way or decides on the state at the event shaded and dated –1 in Fig.5. Then the events in the past of the initial data set dated by –1 and unshaded can now be resolved, thereby extending the original full implication to the left of the initial data set. This process is be called *archaeology*, and for this spacetime could be continued indefinitely into the past. The Theorist can by providing new initial data in this way eventually cover the entire spacetime as the union of an extended initial data set and its full implication.

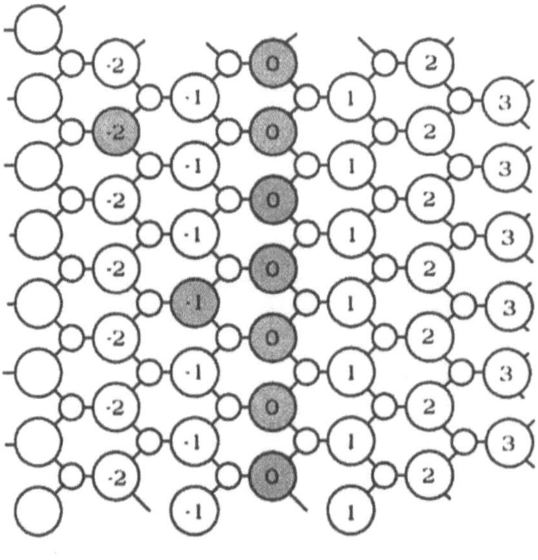

Figure 5. The initial data set is shaded and labelled by zeros. Its first, second and third implications are labelled by 1, 2, and 3 respectively. The past of this initial data set is the set of all events to the left of the zeros. By providing just one extra data event (shaded and labelled –1) the full implication can be extended to include the events unshaded and labelled by –1, and similarly for events labelled by –2.

9. CAUSAL PROPAGATORS AND THE SPEED OF CAUSALITY

Suppose we have been given an initial data set S_0 and have worked out its full implication S_∞. Now pick any event P in S_0 and change its state ψ_P. The consequence of this is to change the states in some events in S_∞, but not necessarily in all events in S_∞. The subset $\mathcal{P}_P(S_0)$ of S consisting of all those events changed by the change in P will be called the *causal propagator* associated with P and S_0. P will be called the *vertex of the propagator*.

We note the following:

1. for any event Q in an initial data set S_0, the causal propagator $\mathcal{P}_Q(S_0)$ lies entirely within the full implication S_∞ of S_0;
2. A causal propagator depends on a vertex *and* on an associated initial data set;
3. A causal propagator divides spacetime into three sets: the vertex, those events which cannot be affected by any change at the vertex, and those events which could be changed. This structure is rather like the lightcone structure in special relativity which separates events into those which are timelike, lightlike, or spacelike relative to the vertex of the lightcone. In our context, we could in some sense talk about the *speed of causality*, analogous to the speed of light in relativity, as the limiting speed with which disturbances could propagate over our spacetime.

Example 3: As an example we give a spacetime lattice of Type A labelled by two integers m, n running from $-\infty$ to $+\infty$. The state space Ω_n^m at each event P_n^m is the set $\{+1,-1\}$ and a state at P_n^m will be denoted by ψ_n^m. The links are given by the equations

$$\psi_n^m \psi_n^{m+1} \psi_n^{m-1} \psi_{n-1}^m \psi_{n+1}^m = 1, \quad -\infty < m, n < \infty.$$

Now choose an initial data set

$$\psi_0^m = \psi_1^m = +1, \quad -\infty < m < \infty.$$

This corresponds to selecting the index m as a spatial coordinate and the index n as a timelike coordinate. The initial data set is then equivalent to specifying the initial values and initial time derivatives of a scalar field on a hyperplane of simultaneity in a two dimensional spacetime.

Because this spacetime dynamics is reversible, the full implication of this initial data set is the entire spacetime minus the initial data set.

Now consider changing the state at m=n=0 from $\psi_0^0 = +1$ to $\psi_0^0 = -1$. In Fig.6 we show a bitmap plot of all those events whose state is changed by the change in the event (0, 0). The structure looks just like a lightcone, with complex fractal-like patterns developing inside the retarded and advanced parts of the lightcone. The speed of causality in this example is evidently unity if we interpret the indices in the manner discussed above.

10. COSMIC TIME

For some spacetime dynamics a global temporal ordering can constructed by assigning an integer to each event as follows. If P is earlier then Q (i.e. P is a partial or complete cause of Q) then some integer p is assigned to P and some integer q to Q such that $p<q$. These integers are called *dates* above.

If it is possible to find a consistent ordering over the whole of S based on the above rule then we may say that a *cosmic time* exists for that spacetime. A cosmic time cannot be constructed for a finite spacetime dynamics if there are no events of degree 1. There are two situations where a cosmic time may be possible: either the spacetime is finite with one or more events of degree one, or the spacetime is infinite. However, these are not sufficient properties to guarantee a cosmic time can be constructed.

It is possible to find spacetime dynamics for which more than one cosmic time pattern can be established.

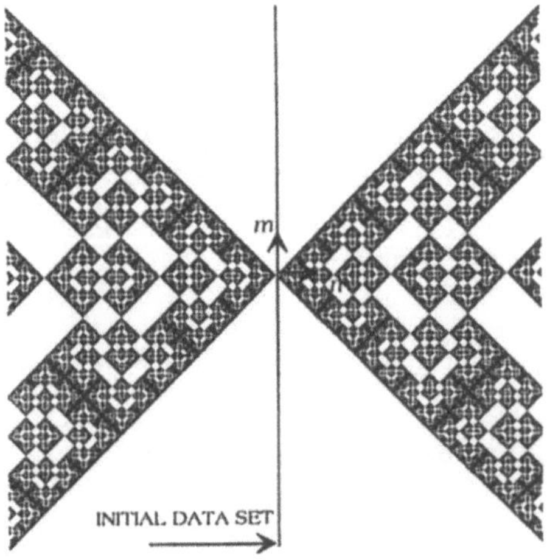

Figure 6. The causal propagator associated with the spacetime dynamics and initial data set in Example 2, with vertex at (0, 0). In this diagram time appears to run left-right.

10.1 Causal (timelike) loops

Fig.7 shows part of a spacetime dynamics containing a closed *causal (timelike) loop*. No cosmic time can be found for such a spacetime. This corresponds to the situation in GR where the existence of a closed timelike loop in a spacetime precludes the possibility of finding a global cosmic time coordinate for that spacetime.

10.2 Spacelike hypersurfaces

In the spacetime depicted in Fig.3 the concept of metric was not introduced. Nevertheless, it is possible to give a definition of a *spacelike (hyper)* surface in this and other spacetime. Whether this corresponds to anything useful depends on the details of the spacetime dynamics.

Definition 5: A *spacelike hypersurface* Σ of a spacetime S is any subset of S which would have a consistent full implication if it were used as an initial data set.

We have already used the term *initial data set* for such subsets. However, not all initial data sets need be consistent. A spacelike hypersurface is just a consistent initial data set.

There may be many spacelike hypersurfaces associated with a given spacetime and they need not all be disjoint. The definition of spacelike hypersurface involves a choice of causation by the Theorist. In general there may be more than one spacelike hypersurface passing through a given event, and it is the Theorist's choice which one to use. The possible non-uniqueness of spacelike hypersurface associated with a given event is desirable, because this is precisely what occurs in Minkowski spacetime, where there is an infinite number of spacelike hypersurfaces passing through any given event, corresponding to hyperplanes of simultaneity in different inertial frames.

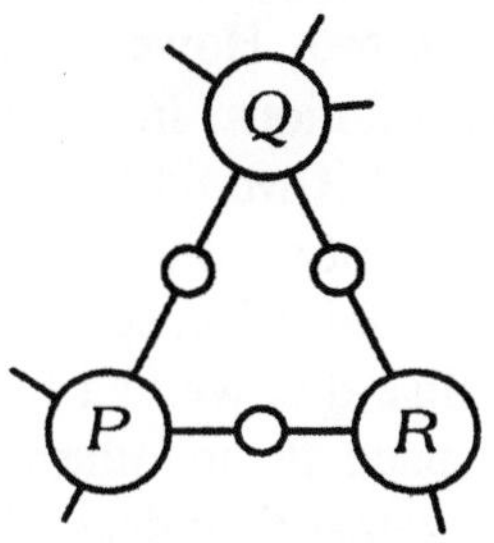

Figure 7. If any spacetime has a causal loop such as *PQR* embedded in it, then no global cosmic time ordering is possible.

11. CAUSAL SETS

Causal sets are sets with some concept of ordering relationship, which makes them suitable for discussions concerning causality[7]. This presupposes some pre-existing temporal structure independent of any dynamical input. This is not a feature of our dynamics, where causal structure emerges only after the Theorist has chosen the initial data set.

12. CONCLUDING REMARKS AND SUMMARY

The PPM model leaves certain fundamental questions unanswered, such as the origin of physiotime and whether process time is a meaningful concept. However, once these are accepted as given and the model used in the right way, then causality and time itself emerge as observer (Theorist) oriented concepts. Many if not all of the phenomena associated with metric theories of spacetime can be recovered, which suggests that further investigation into this approach to spacetime dynamics may prove fruitful.

A good question to ask is: where would Lorentzian causality come from in our approach? The answer is that it is embedded or encoded in the definition of the links. Only when full implications are worked out from given initial data sets would it be noticed that the dynamics itself naturally forces certain patterns to emerge and not others. Only at that stage would the Theorist would recognise an underlying bias in the dynamics in favour of certain more familiar interpretations.

For example, given the continuous time equation

$$\left(\frac{\partial^2}{\partial t^2}-\frac{\partial^2}{\partial x^2}-m^2\right)\varphi(t,x)=0, \tag{8}$$

the notation suggests that t is a time, so we could attempt to solve it with initial data on the hyperplane t=0. However, it would soon emerge that evolution in t gave runaway solutions. In other words, the equation itself would carry the information that it would be wiser to define initial data on the hyperplane x=0. We would not need to invoke the spurious concept of a Lorentzian signature metric embedded in spacetime to discover this. At this stage we might suddenly realise that we had by some chance interchanged the symbols for time and space in an otherwise ordinary Klein-Gordon equation with a real mass. So rather than deal with (8), which behaves like a Klein-Gordon equation with an imaginary mass, we would simply interchange the symbols t, and x and then define initial data on the hyperplane now defined by t=0.

ACKNOWLEDGMENT

The author is grateful to the University of Nottingham for support.

REFERENCES

1. Whitrow G. J., 1980, *The Natural Philosophy of Time*, Clarendon Press, Oxford, 2nd Edition.
2. Howson, A.G., 1972, *A Handbook of Terms Used in Algebra and Analysis*, CUP.
3. 't Hooft G., 1999, *Quantum Gravity as a Dissipative Deterministic System.*
4. Kay B.S., 1998, *Entropy defined, entropy increase and decoherence understood, and some black-hole puzzles solved.*
5. Butterfield J. and Isham C.J., 1999, *Spacetime and the Philosophical Challenge of Quantum Gravity.*
6. Price H., 1997, *Time's Arrow*, Oxford University Press.
7. Meyer D., Bombelli L., Lee J., and Sorkin R., 1999, SpaceTime as a Causal Set, *Phys. Rev. Lett.*, 59 (5), 521-524 (1987); D.P. Rideout and R.D. Sorkin, A Classical Sequential Growth Dynamics for Causal Sets.
8. Frye G., Finkelstein D., and Susskind L., 1974, *Space-time code. V*, Physical Review D, 9 (8) 2231-2236.
9. Rovelli C., 1998, *Strings, loops and others: a critical survey of the present approaches to quantum gravity.*
10. Rovelli C. and Smolin L., 1995, *Spin networks and quantum gravity.*
11. Baez J.C., 1995, *Spin networks in nonperturbative quantum gravity.*
12. Jaroszkiewicz G. and Norton K., 1997, Principles of discrete time mechanics: I. Particle systems, *J. Phys. A: Math. Gen.*, 9 (7) 3115-3144.

Scaling in cosmology and the arrow of time

[1,2]MENAS KAFATOS, [1]SISIR ROY, AND [3]RICHARD AMOROSO
[1]*Center for Earth Observing and Space Research, Institute of Computational Sciences and Informatics*
[2]*Department of Physics, George Mason University, Fairfax, VA 22030-4444, USA*
[3]*The Noetic Institute, Orinda, CA 94563-2502, USA*

1. INTRODUCTION

Cosmological theories and theories of fundamental physics must ultimately not only account for the structure and evolution of the universe, the physics of fundamental interactions but also lead to an understanding of why this particular universe follows the physics that it does. Such theories must lead to an understanding of the values of the fundamental constants themselves. Moreover, the understanding of universe has to utilize experimental data from the present to deduce the state of the universe in distant regions of the past and also account for certain peculiarities or coincidences observed.

The prevalent view today in cosmology is the big bang, inflationary evolutionary model. Although certain nagging problems have remained, e.g. the need to postulate cold, dark matter in amounts much larger than all the observable matter put together, dark matter not detected so far in the laboratory or the recent need to re-introduce the cosmological constant, the big bang cosmology has, nevertheless, achieved impressive results[1].

In this paper we take a different approach than the usual evolutionary picture where the physics itself is assumed invariable. We study some numerical relations among fundamental constants starting from relationships first proposed by Weinberg[2], which turn out to be equivalent to the relations found by Dirac[3], and explore a new scaling hypothesis relating the speed of light c and the scale of the universe R. We then develop an axiomatic approach which results in an *apparent* expanding universe, yielding the same

Studies on the Structure of Time: From Physics to Psycho(path)ology
Edited by Buccheri *et al.*, Kluwer Academic / Plenum Publishers, New York, 2000.

successes as present big bang cosmology but without the need to postulate inflation, cold dark matter, cosmological constant or any of the artificialities of current theory. The "coincidences" of Dirac[3] and Eddington[4] concerning large numbers and ratios of fundamental constants are not to be explained, rather they are accepted and in the process yield a fundamentally different view of the cosmos. The fundamental constants can be assumed to vary with time and this variation leads to an *apparent* expansion of the universe. The variations of the fundamental constants lead to a changing universe, e.g. the number of nucleons varies, etc. The increase of the number of nucleons *appears* to be related to an arrow of time as perceived by an observer in the present universe. Possible implications of this new approach are discussed.

2. FINE TUNING AS IMPLIED BY COSMOLOGICAL OBSERVATIONS

There are a number of observations which must be applied in any cosmological theory which attempts to explain the observed structure of the universe:

a) The universe appears to be quite flat, in other words the density of the universe is close to the so-called closure or critical density,

$$\rho_{crit} = 2\times10^{-29}\left(H_0 \big/ 100\text{km s}^{-1}\text{ Mpc}^{-1}\right)^2 \text{gr cm}^{-3} \tag{1}$$

where H_o is the Hubble constant defined as the apparent rate of expansion with distance, $\dot{R}/R$ and where R is the scale of the universe. In big bang cosmology, this so-called "constant" is actually a function of cosmic time, i.e. it is a variable. Its present-day value seems to be ~ 75 km s^{-1} Mpc^{-1}. The universe appears to be *close* to a flat, Euclidean, Einstein-de Sitter state as indicated from (1), and yet it is still not clear what the geometry of the universe is; *exactly* flat (which would be required by the inflationary scenario); open (yielding a forever-expanding, negatively curved space-time); or closed (yielding a maximum expansion and a positively curved space-time).

b) If one is to assume that the universe followed an inflationary period in the distant past, then the universe must be exactly flat to one part in 10^{50} near the time of big bang. This is the so-called flatness problem: This is such a remarkable requirement that the usual interpretation proposed in the early 80's is that early on, the universe was in an inflationary state, washing out any departures from flatness on time scales of 10^{-35} sec. The inflationary model proposed by Guth and others has been developed in various forms to account for the flatness of the universe and also is proposed to solve the horizon problem, or apparent homogeneity of the 2.73 K black body radiation seen by COBE[5]. The latter problem involves the observation that although the 2.73 K radiation was emitted ~ 10^5 years after the beginning, opposite sides of the sky at that time were out of causal contact, separated by ~ 10^7 light years. Other structures involving large-scale correlations in the universe exist such as very large structures in the distribution of matter[6]. These structures may be progressively hierarchical all the way to the scale of the universe itself.

c) If the universe is indeed flat, observations indicate that baryons (and luminous matter) can only contribute at most ~ 0.05 of the closure density at present. We should ultimately be able to detect the other 90% or more of the matter required to give closure density, presumed to be in the form of cold dark matter[7]. Nevertheless, attempts to detect such exotic matter in the laboratory have, so far, failed. Moreover, the recent realization that the cosmological constant Λ may have to be re-introduced[8] has also led to the probability of Λ itself varying and other similar notions[9]. Without though some direct laboratory verification or overwhelming requirements imposed by particle theory (neither of which presently exists), the nature of dark matter remains elusive. This is clearly a very unsatisfying situation.

d) As we saw, present-day approximate flatness yields to an *exact* flatness in the distant past (this was one of the main reasons why the inflationary scenario was introduced to begin with). The alternative is to accept fine tuning in the universe. In fact, the flatness of the universe is not the only fine tuning. In considering other fundamental observed facts, the universe appears to be extremely fined tuned. It was Eddington[4,10] and Dirac[3] who noticed that certain cosmic "coincidences" occur in nature linking microscopic with macroscopic quantities[11]. A most unusual relationship is the ratio of the electric forces to gravitational forces (this ratio is presumably a constant in an expanding universe where the physics remains constant), or

$$e^2/Gm_e m_p \sim 10^{40} \tag{2}$$

while the ratio of the observable size of the universe to the size of an elementary particle, or

$$R/(e^2/m_e c^2) \sim 10^{40} \tag{3}$$

where in the latter relationship the numerator is changing as the universe expands because the scale of the universe R is constantly changing in an expanding universe.

Dirac formulated the so-called *Large Number Hypothesis* which simply states that the two ratios in (2) and (3) are in fact equal for all practical purposes and postulates that this is not a mere coincidence. Various attempts were made to account for the apparent equality: A possibility that constants such as the gravitational constant may be varying was proposed by Dirac[3] himself and others[12]. Other ratios such as the ratio of an elementary particle to the Planck length,

$$\frac{e^2/m_e c^2}{\left(\hbar G/c^3\right)^{1/2}} \sim 10^{20} \qquad (4)$$

can also be constructed[13] yielding to the conclusion that fine tuning is prevalent in the universe. These relationships may be indicating the existence of some deep, underlying harmonies involving the fundamental constants and linking the microcosm to the macrocosm. Physical theory has not, however, accounted for these in a self-consistent way, waiting perhaps the anticipated unification of all physical forces at the quantum gravity or superstring levels.

e) Evidence has recently been found[14] which seems to be consistent with a time-varying fine structure constant $\alpha = e^2 \hbar / c$. A varying speed of light theory (with $\hbar \; \alpha \; c$) has also been proposed by Albrecht and Magueijo[15]. These two theories correspond to different representations of a varying α in terms of varying dimensional constants. The minimal varying-c theory is of interest because it offers a means of solving the so-called cosmological problems: the horizon, flatness, cosmological constant, entropy and homogeneity problems. Barrow and Magueijo[14] tried to show that there exists a set of duality transformations between these two representations. On the other hand, recent observations of astrophysical events at high redshifts[16,17] can be used to place severe limits on the variation of the speed of light itself ($\Delta c/c$), as well as the photon mass (M_γ).

f) Although other, less traditional ways, such as the Anthropic Principle[18] have been proposed to account for the above fine tuning properties of the universe, there may be another approach involving quantum-like correlations[19].

3. NUMERICAL RELATIONS AND CONCEPT OF SCALING

The critical density of the universe in (1) is defined as

$$\rho_{crit} = \frac{3H_0^2}{8\pi\ G} \tag{5}$$

Let N_p be the number of nucleons in the universe, then

$$m_p = \frac{M}{N_p} = \frac{R\dot{R}^2}{2GN_p} \tag{6}$$

where m_p and M are the mass of the nucleon and mass of the universe, respectively.

Weinberg[2] noticed that one can find a relationship linking the masses of elementary particles to the Hubble constant and other fundamental constants

$$m_\pi \sim \left(\frac{8\hbar^2 H_0}{Gc}\right)^{1/3} \tag{7a}$$

and, correspondingly,

$$m_e \sim \left(\frac{\hbar e^2 H_0}{(8\pi)^3 Gc^2}\right)^{1/3} \tag{7b}$$

where, m_π and m_e are the pion and electron masses, respectively.

These relations can be rewritten as

$$m_p \sim \chi_{p\pi}\left(\frac{8\hbar^2 \dot{R}/R}{Gc}\right)^{1/3} \quad \text{with}\ \chi_{p\pi} = \frac{m_p}{m_\pi}\ \text{and}$$

$$m_p \sim \chi_{pe}\left(\frac{\hbar e^2 \dot{R}/R}{Gc^2 (8\pi)^3}\right)^{1/3} \quad \text{with}\ \chi_{pe} = \frac{m_p}{m_e} \tag{8}$$

From (6) and (7a) one easily gets

$$G^2\hbar^2c^{-1} \sim \chi_{p\pi}^{-3}N_p^{-3}\,\frac{R^4\dot{R}^5}{64} \tag{9}$$

We also have

$$m_p = \chi_{p*}\sqrt{\frac{\hbar C}{G}} \tag{10}$$

where $\chi_{p*} = m_p/m_*$, m_* being the Plank mass and the suffix * indicates Planck quantities. Combining (10) and (6), yields

$$cG\hbar \sim \tfrac{1}{4}N_p^{-2}\chi^{-2}_{p*}R^2\dot{R}^4 \tag{11}$$

Similarly, from (9) and (10)

$$c \sim 2^{2/3}N_p^{-1/3}\chi^{-4/3}_{p*}\chi_{p\pi}\dot{R} \tag{12}$$

The multiplier factor for $\dot{R}$ in (12) is equal to $2^{2/3}N_p^{-1/3}\chi_{p*}^{-4/3}\chi_{p\pi}$, and is ~ 1.

Conversely, if we choose to set $2^{2/3}N_p^{-1/3}\chi_{p*}^{-4/3}\chi_{p\pi} = 1$, one gets the simple relationship linking the speed of light to $\dot{R}$, $c = \dot{R}$ with $N_p \sim 3.7 \times 10^{79}$, which is a good estimate of the number of particles in the current universe. The relationship $c = \dot{R}$ could be interpreted as the Hubble Law $\dot{R} \sim c$, although we emphasize that this is just a relationship and in no way implies that an expansion is indeed taking place.

Similar considerations apply if one chose to work with the relations applying to electrons.

If we start by assuming a heuristic relation

$$c \equiv \dot{R} \tag{13}$$

i.e. the speed of light is *identical* to the rate of change of the scale of the universe, we construct an axiomatic approach equivalent to the Hubble Law. This axiomatic approach can be considered as an alternative approach to the mysterious coincidences of Eddington and Dirac which Weinberg called "so far unexplained... a real though mysterious significance."

It can be further shown that all lengths, such as the Planck length, l_*, the classical electron radius, r_e, etc., are also proportional to

$$R, l_*, r_e, \sim (\ldots)\,R \tag{14}$$

For example,

$$l_{*} \sim (2^{-7/3} N_p^{-1/3} \chi^{5/3}_{p*} \chi^{-2}_{p\pi}) R \tag{15}$$

Similar relations can be found for r_e and r_p where r_e and r_p be the electron and proton radii.

Combining (11) with (14) we obtain

$$G\hbar = \frac{R^2\dot{R}^3}{4} N_p^{-2} \chi^{-2}_{p*} \sim 3.4 \times 10^{-122} R^2 \dot{R}^3 \tag{16}$$

A relationship linking the gravitational and Planck's constant to R and $\dot{R}$, and where the last relationship in (16) holds for the current values of $N_p^{-2} \chi^{-2}_{p*}$ in the universe.

Let us now set the following initial conditions, i.e.,

$$R \to l_{*} \tag{17}$$

$$\dot{R} \to \frac{l_{*}}{t_{*}} = c \tag{18}$$

where l_{*} and t_{*} are the Planck length and Planck time, respectively.

Then $N_p^{-2} \chi^{-2}_{p*}/4 \to 1$ at those initial conditions, while for the present universe the value of this quantity is $\sim 3.4 \times 10^{-122}$.

The limit $N_p \to 1$ indicates that in our model "in the beginning" there was only one bubble-like object or a "cosmic egg"[20]. Moreover, $R \to l_{*}$ and $N_p \to 1$ imply that $\chi_{p*} \to 1$ as well (similarly for all ratios of masses χ's), which in turn indicates that the masses of all particles were equal to each other at these initial conditions. Also, "in the beginning" $R/(e^2/m_ec^2) \sim (e^2/m_ec^2)/(\hbar G/c^3) \sim 1$, rather than the large values of 10^{40} and 10^{20} which these ratios are equal to, respectively, today. "In the beginning" all lengths were equal, all masses were equal and there was only one particle or cosmic egg. Today, these ratios are not unity, there is a very large number of particles in the universe and R is equal to $\sim 10^{28}$ cm. However, scale-invariant relationships such as $c \equiv \dot{R}$; all lengths are proportional to each other, etc. still hold. Israelit and Rosen[20] proposed a cosmological model where the universe emerges also from a small bubble ("cosmic egg") at the bounce point of a de Sitter model filled with a cosmic substrate ("prematter").

In other words, $c \equiv \dot{R}$, at the "initial time" when $N_p \to 1$ and all χ's $\to 1$, and this relationship remains invariant even at the present universe (cf. (12) and (14)). The self-consistency is obtained by calculations for the value of N_p from (12) and (16). This relation is a type of a scaling law and connects the microcosm and the macrocosm.

Now if irrespective (and it is even immaterial) of whether there is expansion of the universe or not, if R itself is changing from the Planck scale to the size of the observable universe, then the fundamental constants like G, $\hbar$, and c also *all* are changing. Note, however, that we cannot deduce the actual variation or the initial value of c and other constants from observations: The relationship $c \equiv \dot{R}$ is not enough to tell us the actual variation or even over "how long" it takes place. It is a scale invariant relationship. If we re-write it as a scale-invariant relationship, $c(t_*)/c(t_0) = \dot{R}(t_*)/\dot{R}(t_0)$ where t_* and t_0 could be conveniently taken as the Planck time and the present "age" of the universe, then this relationship is not enough to give us the evolution of $\dot{R}$ or even the values of t_* and t_0. Hence it cannot tell us how c itself is varying or even if it is varying. If we wanted to insist that c *is constant*, then all the other "constants" like G and $\hbar$ *are really constant* as well. But if c is not constant, then all the other "constants" are varying as well. In both cases, however, the number of particles is changing, the ratios of masses are changing and the ratios of scales or lengths are also changing. An arrow of time *could*, therefore, be introduced. In this picture, invariant relationships hold and from unity, there is evolution into diversity. One cannot though conclude how the variations are taking place, over what timescales they are taking place or even how old the universe is. The universe could be 10^{10} years old or 5×10^{-44} sec (the Planck time) old, or any time in between. *Time is strictly a parameter* that can be introduced in the scale-invariant relationships. It has no meaning by itself. The universe *appears to be evolving* as the number of particles and ratios are varying.

4. DISCUSSION AND CONCLUSIONS

The existence of horizons of knowledge in cosmology, indicate that as a horizon is approached, ambiguity as to a unique view of the universe sets in. It was precisely these circumstances that apply at the quantum level, requiring that complementary constructs be employed[21]. At the *initial time*, which could be conveniently taken as the Planck time, if we set the conditions like $c = \dot{R}$, as proposed in this paper, we can axiomatize the numerical relations connecting the microcosm and the macrocosm. One then has scale-invariant relationships. During the *evolutionary* process of the universe, the fundamental constants are changing or they may be constant. In the former case, we don't know *how* or even over *what* timescales they are changing. In the latter case, one gets the usual evolutionary universe. This is a clear case where complementarity applies.

In other words, as N_p is changing from the initial value of 1 (unity) to the present large value of $\sim 10^{80}$ (diversity), more particles are created as R and all length scales as well as all masses are changing. This could be interpreted by an observer as an "expansion of the universe". An observer, who is inside the universe will perceive an "arrow of time" and an "evolving universe". But equivalently, as the "constants" change (in contrast to previous works, they would *all* have to be changing), or even if they are truly constant, there *appears* to be an evolution. As $N_p \rightarrow 10^{80}$, the present number of the nucleons in the universe, the fundamental "constants" achieve their present values.

To recapitulate, the arrow of time can be related to a kind of complementarily between two constructs, i.e., the fundamental "constants" *are truly constant*, on the one hand; and the fundamental "constants" *are changing*, on the other hand.

In summary, we found that by adopting Weinberg's relationship (which can be shown to be equivalent to Dirac's relationships (2) and (3) when the latter are equated to each other), we can obtain a relationship linking the speed of light *c* to the rate of change of the scale of the universe. In fact, the proportionality factor is ~ 1 if one substitutes for values of fundamental quantities like the present number of particles in the universe, etc. The next step assumes that the relationship linking c and R is an identity, i.e. $c \equiv \dot{R}$ (for example, at the Planck time, one observes that this relationship still holds if the ratios of all masses $\rightarrow 1$ and the number of particles also $\rightarrow 1$). As such, it is possible (but not necessary) to state that *all* the fundamental constants are changing and not just one of them as was assumed in past works. It is interesting that, recently, the possibility of the cosmological constant Λ itself changing[9] has been suggested. As such, what we are suggesting here as a framework for the universe is a natural extension of previous ideas. Therefore, as N_p changes from an initial value of 1 to the present value of 10^{80} ($1 \rightarrow 10^{80}$), the universe would be appearing to be evolving to an observer inside it or an arrow of time is introduced. Finally, the outcomes of this prescription are not just that an arrow of time is introduced and the mysterious coincidences of Dirac and Eddington now can be understood as scale-invariant relationships linking the microcosm to the macrocosm; but in addition, all scales are linked to each other and what one calls, e.g. *fundamental length,* etc. is purely a convention. In the same way, time itself is not as fundamental as the scale-invariant relationships linking the microcosm to the macrocosm.

REFERENCES

1. Silk J., 1989, *The Big Bang*, W.H. Freeman, New York.
2. Weinberg S., 1972, *Gravitation and Cosmology: Principles and Applications of the General Theory of Relativity*, Wiley, New York.
3. Dirac P.A.M., 1937, *Nature*, **139**, 323.
4. Eddington A.S., 1931, *M.N.R.A.S.*, **91**, 412.
5. Smoot G.F., 1996, in *Examining the Big Bang and Diffuse Background Radiations*, Kafatos and Kondo (eds.), Kluwer Academic Publishers, Dordrecht, 31.
6. Geller M.J., Huchra J., 1989, *Science*, **246**, 897.
7. Novikov I., 1996, in *Examining the Big Bang and Diffuse Background Radiations*, Kafatos and Kondo (eds.), 289, Kluwer Academic Publishers, Dordrecht.
8. Peebles P.J.E., 1998, "Is Cosmology Solved?", *The Nature of the Universe Debate: Cosmology Solved?*, Smithsonian Institution.
9. Glanz J., 1998, *Science*, **282**, 2156.
10. Eddington A.S., 1939 *The Philosophy of Physical Science*, Cambridge University Press, Cambridge.
11. Kafatos M., 1989, in *Bell's Theorem, Quantum Theory and Conceptions of the Universe*, Kafatos (ed.), 195, Kluwer Academic Publishers, Dordrecht.
12. Dyson F.J., 1972, *Aspects of Quantum Theory*, Salam and Wigner (eds.), Cambridge University Press, Cambridge.
13. Harrison E. R., 1981, *Cosmology: The Science of the Universe*, 329, Cambridge University Press, Cambridge.
14. Barrow J.D., Magueijo J., 1998, "Varying α Theories and Solutions to the Cosmological Problems," *Astro-ph*/ 9811072.
15. Albrecht A., Magueijo J., 1998, *"A time-varying speed of light as a solution to cosmological puzzles"* (preprint).
16. Schaefer E.B., 1998, *Severe Limits on Variations of the Speed of Light with Frequency*, Yale University (preprint).
17. Amelio-Camella G., et al., 1998, *Nature*, **393**, 763.
18. Barrow, J.D., Tipler F.J., 1986, *The Anthropic Cosmological Principle*, Clarendon Press, Oxford.
19. Roy S., Kafatos M., 1998, *"Bell-type Correlations and Large Scale Structure of the Universe"*, (preprint).
20. Israelit M., Rosen N., 1989, *ApJ*, **342**, 25.
21. Bohr N., 1961, *Atomic Theory and the Description of Nature*, Cambridge University Press, Cambridge.
22. Aspect A., Grangier P., Roger G., 1982, *Phys. Rev. Lett.*, **49**, 91.
23. Bekenstein D.J., 1982, *Phys. Rev. D*, **25**, 1527.
24. Bell J.S., 1964, *Physics*, **1**, 195.
25. Blome J.H., Priester W., 1991, *Astr.and Astrphys.*, **250**, 43.
26. Guth A., 1981, *Phys. Rev. D.*, **23**, 347.
27. Kafatos M., Nadeau R., 1990, *The Conscious Universe: Part and Whole in Modern Physical Theory*, Springer-Verlag, New York.

The Arrow of Time

First direct observation of Time-Reversal Non-Invariance

DANILO ZAVRTANIK, for the CPLEAR Collaboration*
Nova Gorica Polytechnic and J. Stefan Institute, Ljubljana, Slovenia

1. INTRODUCTION

The concept of time and its implications in physics and cosmology are discussed in many textbooks. A "brief history of time" has been told in a recent book by S. Hawking[1] and his basic ideas on the arrow of time are briefly summarized here as an introduction.

There are at least three different definitions of the arrow of time in the macrocosm. First, we have the thermodynamic arrow of time. This is a consequence of the second low of thermodynamics stating that in any closed system entropy, or disorder, always increases with time. Thus, the arrow of

* A. Angelopoulos[a], A. Apostolakis[a], E. Aslanides[k], G. Backenstoss[b], P. Bargassa[m], O. Behnke[r], A. Benelli[b], V. Bertin[k], F. Blanc[g,m], P. Bloch[d], P. Carlson[o], M. Carroll[i], E. Cawley[i], M.B. Chertok[c], M. Danielsson[o], M. Dejardin[n], J. Derre[n], A. Ealet[k], C. Eleftheriadis[p], W. Fetscher[r], M. Fidecaro[d], A. Filipcic[j], D. Francis[c], J. Fry[i], E. Gabathuler[i], R. Gamet[i], H.J. Gerber[r], A. Go[d], A. Haselden[i], P.J. Hayman[i], F. Henry-Couannier[k], R.W. Hollander[f], K. Jon-And[o], P.R. Kettle[m], P. Kokkas[b], R. Kreuger[f], R. Le Gac[k], F. Leimgruber[b], I. Mandic[j], N. Manthos[h], G. Marel[n], M. Mikuz[j], J. Miller[c], F. Montanet[k], A. Muller[n], T. Nakada[m], B. Pagels[r], I. Papadopoulos[p], P. Pavlopoulos[b], G. Polivka[b], R. Rickenbach[b], B.L. Roberts[c], T. Ruf[d], L. Sakeliou[a], M. Schäfer[r], L.A. Schaller[g], T. Schietinger[b], A. Schopper[d], L. Tauscher[b], C. Thibault[l], F. Touchard[k], C. Touramanis[i], C.W.E. Van Eijk[f], S. Vlachos[b], P. Weber[r], O. Wigger[m], M. Wolter[r], D. Zavrtanik[j], and D. Zimmerman[c].

[a]University of Athens, Greece, [b]University of Basle, Switzerland, [c]Boston University, USA, [d]CERN, Geneva, Switzerland, [e]LIP and University of Coimbra, Portugal, [f]Delft University of Technology, Netherlands, [g]University of Fribourg,Switzerland, [h]University of Ioannina, Greece, [i]University of Liverpool, UK, [j]J. Stefan Inst. and Phys. Dep., University of Ljubljana, Slovenia, [k]CPPM, IN2P3 CNRS et Université d'Aix Marseille II, France, [l]CSNSM, IN2P3 CNRS, Orsay, France, [m]Paul Scherrer Institut (PSI), Villigen, Switzerland, [n]CEA, DSM/DAPNIA, CE Saclay, France, [o]Royal Institute of Technology, Stockholm, Sweden, [p]University of Thessaloniki, Greece, [r]ETH IPP Zürich, Switzerland.

time points towards the direction in which entropy increases. Then, there is the psychological arrow of time, the direction in which we feel time passes. This definition is also connected to the fact that we remember the past but not the future. Finally, we have the cosmological arrow of time. This is the direction of time in which the Universe is expanding. All the three arrows of time point at the same direction. One can argue that psychological arrow of time is determined by thermodynamic arrow. Thus, the two definitions are not independent and they necessarily point to the same direction. However, thermodynamic and cosmological arrows of time are well defined. They point to the same direction although it may not be so for the whole history of our Universe.

Although there is a big difference between the forward and backward directions of time in ordinary life, the fundamental classical laws of physics discovered by Galilei, Newton and Einstein do not distinguish between the past and the future. In other words, they appear to be symmetric in time. Thus, one might expect that the thermodynamic arrow of time would be absent in the cold microcosm of elementary particles. Therefore, understanding the nature of time-reversal symmetry and its violation appears essential for a full understanding of the concept of time.

Where to search for time asymmetric processes in the microcosm? To the present knowledge, we are living in a CPT conserving world. On the other hand, CP violation in weak interactions was discovered almost 35 years ago. As a consequence we expect weak processes to be time asymmetric. The neutral-kaon system proved to be a unique experimental environment to test discrete symmetries. Yet, it is the only system in nature where the violation of CP symmetry has been observed. In addition, experimental observation of interference effects allows direct measurements of T violation and CPT tests with a precision unachievable in other systems.

No direct observation of time-reversal non-invariance has never been performed up to recently. Following an original proposal of Kabir[2] and Aharony[3], a direct evidence for T violation can be obtained by comparing the transformation rates of $\overline{\mathrm{K}}^0$ to K^0 and K^0 to $\overline{\mathrm{K}}^0$ which are two T-conjugate processes. This test was performed by CPLEAR Collaboration at CERN and will be reported in more details in the present article.

2. NEUTRAL-KAON SYSTEM

The time evolution of a state $|\psi\rangle = a(t)|\mathrm{K}^0\rangle + \bar{a}(t)|\overline{\mathrm{K}}^0\rangle$ is given by

$$i\frac{d}{dt}\begin{pmatrix} a(t) \\ \bar{a}(t) \end{pmatrix} = \hat{H}\begin{pmatrix} a(t) \\ \bar{a}(t) \end{pmatrix} \tag{1}$$

where $\hat{H}$ is a 2×2 matrix. As the weak interaction allows second order weak transitions ($K^0 \leftrightarrow \overline{K}^0$) and kaon decays, the matrix is non-diagonal and complex[4] $\hat{H} = M - \frac{i}{2}\Gamma$. Here M is the mass and Γ the decay matrix, both of them being Hermitian.

K_L and K_S are eigenstates of $\hat{H}$. They have different lifetimes ($\tau_S = (89.34 \pm 0.08)$ ps and $\tau_L = (51.7 \pm 0.4)$ ns $\approx$ 580 τ_S) and masses ($\Delta m = m_L - m_S = (530.1 \pm 1.4) \times 10^7\, \hbar/\mathrm{s} = (3.489 \pm 0.009) \times 10^{-12}$ MeV)[5].

Invariance under time-reversal requires equality of magnitudes of off-diagonal matrix elements $\hat{H}_{12}$. Together with M and Γ being Hermitian this is equivalent to colinearity of M_{12} and Γ_{12}.

$$\begin{aligned} &\left|M_{12} - \tfrac{i}{2}\Gamma_{12}\right| = \left|M_{21} - \tfrac{i}{2}\Gamma_{21}\right| \Leftrightarrow \\ &\Leftrightarrow \arg(\Gamma_{12}) - \arg(M_{12}) = n \cdot \pi \end{aligned} \quad \mathcal{T}\text{invariance} \tag{2}$$

Invariance under CPT implies equality of K^0 and $\overline{K}^0$ masses and lifetimes.

$$\begin{aligned} M_{11} &= M_{22} \\ \Gamma_{11} &= \Gamma_{22} \end{aligned} \quad \mathcal{CPT}\text{invariance} \tag{3}$$

Violation of either T or CPT results in violation of CP. Experimentally, CP violation is observed to be small (of the order of 10^{-3}) and while $\Gamma_S \gg \Gamma_L$, the mass $m_S < m_L$. Thus, violation of T in mixing gives rise to a single small T-violation parameter $\delta\varphi$

$$\arg(\Gamma_{12}) - \arg(M_{12}) = \pi - \delta\varphi. \tag{4}$$

A convenient T violation parameter is ε_T, connected to $\delta\varphi$ by

$$\varepsilon_T = \frac{2i}{\Delta\Gamma}\,\frac{\left|\hat{H}_{12}\right|^2 - \left|\hat{H}_{21}\right|^2}{2(\Lambda_L - \Lambda_S)} = \frac{\Delta m}{2}\cdot\frac{i\Delta m + \Delta\Gamma/2}{\Delta m^2 + (\Delta\Gamma/2)^2}\cdot\delta\varphi, \tag{5}$$

where $\Gamma_{L,S}=m_{L,S}-i\Gamma_{L,S}/2$ and $\Delta\Gamma=\Gamma_S-\Gamma_L$ ε_T is a phase-convention related to ε by $\varepsilon=\varepsilon_T+i\phi_\Gamma/2$ with ϕ_Γ being the phase of Γ_{12} [6]. CPT violation in mixing results in a complex parameter δ_{CPT}

$$\begin{aligned}\delta_{CPT} &= \frac{\hat{H}_{22}-\hat{H}_{11}}{2(\Lambda_L-\Lambda_S)} \\ &= \frac{i}{2}\left[(M_{11}-M_{22})-\frac{i}{2}(\Gamma_{11}-\Gamma_{22})\right]\cdot\frac{i\Delta m+\Delta\Gamma/2}{\Delta m^2+(\Delta\Gamma/2)^2}.\end{aligned} \tag{6}$$

The argument of ε_T is the superweak angle ϕ_{SW}=arctan($2\Delta m/\Delta\Gamma$)$\approx\pi/4$. In δ_{CPT}, the component due to K^0 and $\overline{K}^0$ lifetime difference is parallel to ε_T and the one due to K^0 and $\overline{K}^0$ mass difference perpendicular to it.

3. CPLEAR EXPERIMENT

In the CPLEAR experiment, initial K^0 and $\overline{K}^0$ are produced concurrently via the annihilations of antiprotons at rest into $K^0K^-\pi^+$ and $\overline{K}^0K^+\pi^-$. Both reactions occur at a branching ratio of $\approx 2\times10^{-3}$. The antiprotons are delivered by the low-energy antiproton ring LEAR at CERN and are stopped in a gaseous hydrogen target in the center of the CPLEAR detector. The tagging of opposite strangeness states at production time maximizes the interference effects to be observed in K^0, $\overline{K}^0$ decays.

The experimental apparatus[7] is shown in Fig. 1. Ten chamber layers (2 Proportional Chambers, 6 Drift Chambers, 2 Streamer Tubes) were used to trace charged particles resulting from annihilation and neutral-kaon decays. A 32-segment sandwich of scintillator-Cherenkov-scintillator detectors provided particle identification (kaons/pions/electrons). Photons were detected by a 18-layer fine-grain streamer tube/lead sampling calorimeter. Signals from all detectors were processed in a multilevel trigger, providing a rejection factor of over 1000 and allowing the detector to operate at a $\overline{p}$ rate of 1 MHz. The material in the decay region up to the streamer tubes was minimized by using a gas target with mylar-kevlar walls and innovative low-mass chamber construction, thus reducing regeneration effects of neutral kaons. In 1995 a proportional chamber was added at 1.7 cm radius to improve the trigger and tracking capabilities. Data in 1996 were taken with a carbon regenerator to measure regeneration amplitudes[8].

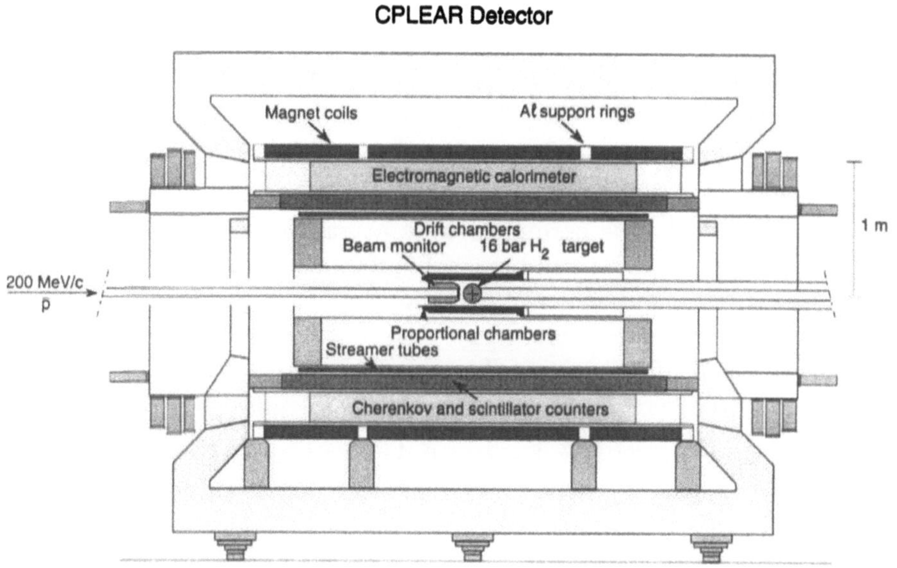

Figure 1. View of the CPLEAR detector.

In total, about 100 million K^0 and $\overline{K}^0$ decays were reconstructed. The presented results refer to the analysis of the complete data-set of 70 M K^0, $\overline{K}^0 \to \pi^+\pi^-$ decays with $t > 1\,\tau_S$, 1.3 M K^0, $\overline{K}^0 \to e\pi\nu$ decays, 0.5 M K^0, $\overline{K}^0 \to \pi^+\pi^-\pi^0$ decays, 2 M K^0, $\overline{K}^0 \to \pi^0\pi^0$ decays and 17 k K^0, $\overline{K}^0 \to \pi^0\pi^0\pi^0$ decays, extracted from about 50000 data tapes with 100000 events each.

4. T VIOLATION

Since weak interactions do not conserve strangeness, a K^0 meson can transform into a $\overline{K}^0$ and vice versa. Time-reversal invariance would require all details of the second process to be deducible from the first. In particular, the probability ($\mathcal{P}$) that a $K^0(t{=}0)$ is observed as a $\overline{K}^0$ at time τ should be equal to the probability that a $\overline{K}^0(t{=}0)$ is observed as a K^0 at the same time τ [2]. Any difference between this two probabilities is a signal for T violation and can be measured through time-reversal asymmetry

$$\frac{\mathcal{P}(\overline{K}^0 \to K^0) - \mathcal{P}(K^0 \to \overline{K}^0)}{\mathcal{P}(\overline{K}^0 \to K^0) + \mathcal{P}(K^0 \to \overline{K}^0)}. \tag{7}$$

Experimentally this requires knowledge of the strangeness of the neutral kaon at its production and decay. The initial strangeness of the neutral kaon is tagged by the charge of the accompanying charged kaon at the production time. To tag the strangeness of the kaon at the decay time we use semileptonic decays $K^0 \rightarrow e^+\pi^-\nu$ and $\overline{K}^0 \rightarrow e^-\pi^+\overline{\nu}$. A measured positive lepton charge is associated to a K^0 and negative lepton charge to a $\overline{K}^0$. Electron identification is performed by a neural-network based technique, making use of energy loss and time-of-flight in the scintillators and number of photoelectrons in the Cherenkov detector.

There are four measurable decay rates R, labeled by the initial kaon strangeness and the final electron charge:

$$R^+(\tau) = R(K^0_{t=0} \rightarrow e^+\pi^-\nu_{t=\tau}) \qquad R^-(\tau) = R(K^0_{t=0} \rightarrow e^-\pi^-\nu_{t=\tau})$$
$$\overline{R}^-(\tau) = R(\overline{K}^0_{t=0} \rightarrow e^-\pi^+\overline{\nu}_{t=\tau}) \qquad \overline{R}^+(\tau) = R(\overline{K}^0_{t=0} \rightarrow e^+\pi^-\nu_{t=\tau}).$$

The decay amplitudes are parametrized with quantities which have well defined symmetry properties under T, CP and CPT symmetries,

$$\begin{aligned} &\left\langle l^+\pi^-\nu \,|\, T \,|\, K^0 \right\rangle = a + b, \quad \left\langle l^-\pi^+\overline{\nu} \,|\, T \,|\, \overline{K}^0 \right\rangle = a^* - b^*, \\ &\left\langle l^-\pi^+\overline{\nu} \,|\, T \,|\, K^0 \right\rangle = c + d, \quad \left\langle l^+\pi^-\nu \,|\, T \,|\, \overline{K}^0 \right\rangle = c^* - d^*, \end{aligned} \tag{8}$$

where l denotes an electron or a muon and T the decay transition matrix. $\mathcal{Re}(a)$ is CP, T and CPT symmetric, all the imaginary parts are T violating, b and, equivalently, $y=-b/a$ violate CPT. The last two transitions (c and d) violate the $\Delta S=\Delta Q$ rule, whose validity is parametrized by $x=(c^*-d^*)/(a+b)$ and $\overline{x}=(c^*+d^*)/(a-b)$, or $x_+=(x+\overline{x})/2$ and $x_-=(x-\overline{x})/2$, with x_- also violating CPT and $x_+\leq 10^{-6}$ in the Standard Model.

The asymmetry

$$A_T = \frac{R(\overline{K}^0 \rightarrow K^0) - R(K^0 \rightarrow \overline{K}^0)}{R(\overline{K}^0 \rightarrow K^0) + R(K^0 \rightarrow \overline{K}^0)} \tag{9}$$

represents a direct test of T reversal through detailed balance[2].

Relative tagging efficiencies at the production and decay time differ due to interactions in detector materials. The efficiency correction for the initial strangeness tag is parametrized by the efficiency ratio $\xi=\varepsilon(K^+\pi^-)/\varepsilon(K^-\pi^+)$ while the correction for the relative detection efficiency of $e^+\pi^-$ versus $e^-\pi^+$ is applied through $\eta=\varepsilon(\pi^+e^-)/\varepsilon(\pi^-e^+)$. With the measured number of events N corrected for η and ξ, the time-reversal asymmetry reads

$$A_T^{exp} = \frac{\eta \overline{N}^+ - \xi N^-}{\eta \overline{N}^+ + \xi N^-}. \tag{10}$$

Calibration data were used to obtain η in bins of (p_e, p_π). Its average value was $\langle\eta\rangle = 1.014 \pm 0.002$. The parameter ξ was taken from $\pi^+\pi^-$ decays through $\alpha_{2\pi} = [1+4\mathcal{Re}(\varepsilon_T - \delta_{CPT})]\times\xi$, where $\alpha_{2\pi}$ is obtained by using the measured (unweighted) number of $K^0, \overline{K}^0 \rightarrow \pi^+\pi^-$ decays in the decay-time interval 1-4 τ_S [9]. External information on $\mathcal{Re}(\varepsilon_T - \delta_{CPT})$ was taken from the semileptonic charge asymmetry[5]

$$\delta_l = 2\mathcal{Re}(\varepsilon_T - \delta_{CPT}) - 2\mathcal{Re}(x_- + y) = (3.27 \pm 0.12)\times 10^{-3},$$

which is equal to $2\mathcal{Re}(\varepsilon_T - \delta_{CPT})$ in the limit of CPT symmetry in semileptonic decays, and an event-by-event correction was applied according to primary pair kinematics. The average value of ξ is $\langle\xi\rangle = 1.12023 \pm 0.00043$.

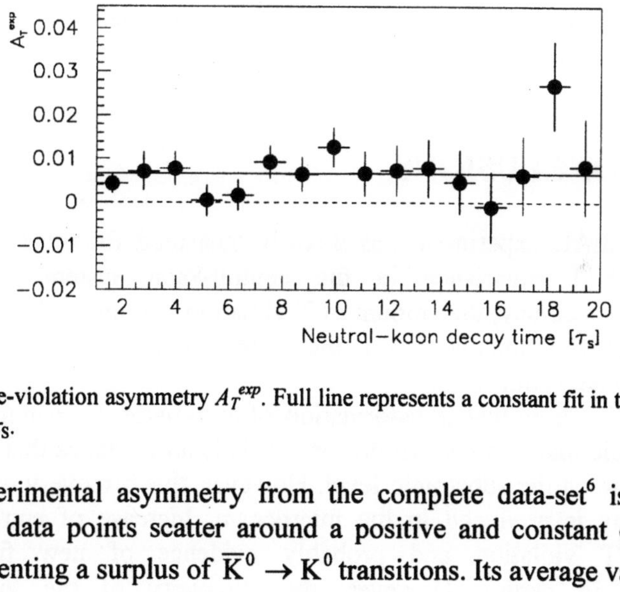

Figure 2. Time-violation asymmetry A_T^{exp}. Full line represents a constant fit in the decay-time interval 1-20 τ_S.

The experimental asymmetry from the complete data-set[6] is shown in Fig. 2. The data points scatter around a positive and constant offset from zero, representing a surplus of $\overline{K}^0 \rightarrow K^0$ transitions. Its average value

$$\left\langle A_T^{exp} \right\rangle_{(1-20)\tau_S} = (6.6 \pm 1.3_{stat} \pm 1.0_{syst})\times 10^{-3}$$

represents the first direct measurement of time-reversal violation.

To clarify the eventual influence of $\Delta Q \neq \Delta S$ or CPT-violating transitions one has to consider the phenomenological expression for A_T^{exp}

$$A_T^{exp} = 4(\mathcal{R}e(\varepsilon_T - y) - \mathcal{R}e(x_-)) + \\ +2\frac{\mathcal{R}e(x_-)\left(e^{-\frac{1}{2}\Delta m\tau} - \cos(\Delta m\tau) - \mathcal{I}m(x_+)\sin(\Delta m\tau)\right)}{\cosh(\frac{1}{2}\Delta\Gamma\tau) - \cos(\Delta m\tau)}. \tag{11}$$

With the transition $\alpha_{2\pi}\rightarrow\xi$ using δ_l, an additional term $-2\mathcal{R}e(y+x_-)$ entered into A_T^{exp}. In the long lifetime limit Eq. (11) reads

$$A_T^{exp} \xrightarrow{\tau>>\tau_s} 4\mathcal{R}e(\varepsilon_T) - 4\mathcal{R}e(y + x_-). \tag{12}$$

By using the Bell-Steinberger relation CPLEAR has determined[10] that

$$\mathcal{R}e(y + x_-) = (-0.2 \pm 0.3) \times 10^{-3}.$$

This result, based on the sole assumption of unitarity, confirms the interpretation of A_T^{exp} as the first direct measurement of time reversal violation.

5. CONCLUSIONS

The CPLEAR experiment has directly measured for the first time the violation of T invariance in the neutral-kaon system. Within the experimental accuracy, the amount of T violation is found to be equal to the amount of CP violation observed in the neutral-kaon system and is consistent with CPT conservation.

Our results show that transformation of anti-matter to matter is slightly more probable than the reverse process. This is an evidence that there is an arrow of time on the subatomic level. However, this discovery does not tell us if entropy is produced in the microcosm. Increase of entropy would require CPT violation and probably evidence of new fundamental interactions in nature. However, the discovery of the violation of time-reversal invariance might have far-reaching consequences for our understanding of the cosmos.

REFERENCES

1. Hawking S.W., 1996, *A brief history of time: from big bang to black holes*, Bantam Books, Toronto.

2. Kabir P., 1970, *Phys. Rev.* **D2**, 540.
3. Aharony A., 1970, *Lett. Nuovo Cimento* **3** 791.
4. Weisskopf V., Wigner E., 1930, *Z.Phys.* **63** 54.
5. Caso C. et al., 1998, Particle Data Group, *Eur. Phys. J.* **C3**, 1.
6. Angelopoulos A. et al., 1998, CPLEAR Collab., *Phys. Lett.* **B444**, 43.
7. Adler R. et al., 1996, CPLEAR Collab., *Nucl. Instr. Meth.* **A379** 76.
8. Angelopoulos A. et al., 1997, CPLEAR Collab., *Phys. Lett.* **B413** 422.
9. Angelopoulos A. et al., 1998, CPLEAR Collab., *Phys. Lett.* **B444** 52.
10. Apostolakis A. et al., 1999, CPLEAR Collab., *Phys. Lett.* **B456**, 297.

3.

REAL OR ILLUSIVE TIME?

The Parameters of Temporal Correspondence in a Continuous State Conscious Universe

RICHARD L. AMOROSO
The Noetic Institute - Physics Lab
120 Village Square, MS 49 Orinda, CA 94563-2502 USA

"And as imagination bodies forth the form of things unknown, the poet's pen turns them into shapes, and gives to airy nothing a local habitation and a name". *W. Shakespeare*
"The distinction between past, present, and future is an illusion...". *A. Einstein*

1. INTRODUCTION

Clock time can appear absolute in the classical Newtonian sense or relativistic for accelerated observers. The perception of time is coupled to dynamical processes often associated with entropic flow; but entropic time does not correlate with clock time. And clock time only correlates with psychological time for certain states of consciousness. If a process is harmonic or occurs at a microscopic level where the laws of physics are temporally symmetric all conception of time can be lost; thus the nature of time has maintained itself as a dilemma for millennia.

In this chapter Einstein's notion of eternal time in a static universe will be extended to develop a framework for correspondences that unify all aspects of time within a new cosmological model called the continuous state conscious universe (CSCU)[1]. Within the CSCU framework human existence is composed of a two-fold complementarity; a Cartesian type body/mind dualism comprised of a *res extensa*, and *res cogitans*[2]. The component associated with the body (obeying Fermi-Dirac statistics) is governed by classical dynamics, the rules of general relativity, and imbedded in a curved 3 (4)-D Minkowski-Riemann spacetime. The latter component (obeying Bose-Einstein statistics) is governed by the laws of special relativity and imbedded in an 'eternal' locally flat 11 (12)-D higher dimensional space that

Studies on the Structure of Time: From Physics to Psycho(path)ology
Edited by Buccheri *et al.*, Kluwer Academic / Plenum Publishers, New York, 2000.

mediates a *vis vitae* or *elan vital*[3] through action of a teleological principal inherent to the unified noetic field originating nonlocally in a 'supralocal' domain.

There is not simply 'one nonlocality' as most physicists assume. Nonlocality (NL) has a manifold structure comprised of a NL I - Spacial Nonlocality, NL II - Temporal Nonlocality & NL III - Unitarity; each component of which has a different structural-phenomenology[4]. The putative supralocality is coupled to NL III except that it has domains that are not causally connected to our 10 (11)-D 'virtual reality'. (See Figure 8)

A complete description of this structural-phenomenology can not be done adequately from within the domain of the current standard models; and demands elaboration from the point of view of extended physical theory and embracing currently unpopular philosophical constructs like vitalism and teleology. An understanding of the nature of time is an essential part of the extended theory[5] and will shed additional light on the nature of gravitation, including the contrast between determinism and free will. This is why a description of the CSCU is a required precursor. Within the CSCU context it is possible to present a comprehensive model of the nature of the origin of time, life and mind for the first time. In addition this ultimately entails describing vacuum scale gravitational theory and completing quantum theory by integrating non-collapse evolution of the wave function by introducing an additional weak causal principal of consciousness which the current Copenhagen interpretation does not include.

Time is inexorably connected with space in the entity spacetime, which is not fundamental, but in which "We" and the properties of all matter we perceive is imbedded. It is the properties of this matter, that our awareness is imbedded in also, with which we define the measuring rods that are the fundamental basis of all physics[6]. To understand the nature or foundations of time we must be able to comprehend more fully the nature of space. As Einstein obviated the absolute space of Newtonian mechanics; we must now obviate the classical basis of measurement used up to this point in the history of empiricism, and in the process obviate and reformulate the whole fundamental basis of physics itself. It is mandatory that a new scientific methodology be devised to investigate the 'absolute noumenon of existence' that resides behind the 'facade of phenomenological reality'. If scientists are finally ready to open this Pandora's box?

2. REALITY AS ILLUSION

'While walking across the campus lawn toward the UC Berkeley physics library absorbed in thought, an intrusive sound caused me to raise my head.

In front of me was a pick-up truck; I *heard* the engine running. Two steps later, with a new parallax, the sound separated into components. A few meters away toward the right was a drain making the gurgling sound of underground water running into it. After another step a second sound component separated. Rectilinearly 500 meters away, up a few stories in a construction site, the wine of an electric drill emanated. I looked at the truck again; the engine was not running. The two sounds combined, plus the fixation of my sight on the truck, produced the completely accepted "real" but illusory experience in my "minds eye" of a truck's engine running!

It is suggested that our fundamental awareness is comprised of base states[2,7,8,9] imbedded in spacetime that evanesce into our perceived reality in a manner similar to the above serendipitous experience. This concept was described anciently in Plato's 'analogy of the cave' as shown in Figure 1. More recently: "Time and space are modes by which we think and not conditions in which we exist." - Albert Einstein, 1941. The physical nature of our mind and apprehended reality are a summation or superposition - a complex hyperdimensional 'standing wave' of spacetime and consciousness base state components comprising both our temporal existence and physical nature of our temporal awareness. This premise is taken as the starting point for our discourse.

Plato's Cave/Einstein's Mode of Though

Figure 1. The 'facade' of sensory reality. For millennia scientists, philosophers and theologians have considered reality as a perceptual illusion; a limited phenomenology of observation imbedding us in the material comprising it. Plato said 'that reality is as if we are in a cave chained up facing a wall viewing events by the light of a fire projected from behind never knowing the true nature of existence. If we were released to turn toward the light at first we would be blinded by its brightness still having our perception clouded.' Finally scientists seem ready to solve this paradox - that the basic assumptions and all the empirical measurements forming the foundation of physical theory are not are not fundamental, but based on an illusion of perceptual phenomenology that our existence is structurally imbedded in. This requires new Physics and Cosmology defining time and how our perception is involved in the parameters of spacetime.

But obviating all of physics is not to discard it; the falsification of Newtonian mechanics for example merely relegated it to its absolute domain. By obviating 'reality' science defines the classical limit and prepares to describe a new framework redefining the physical world and creating the formalism and additional experimental methods needed to study the 'noumenon' of the universe beyond the 'illusion'. In this chapter fundamental parameters of the cosmology that our awareness is imbedded in are outlined. It appears that it will not be possible to comprehend the nature of space, time or mind until this task is adequately underway.

3. PHILOSOPHY OF MIND - THE HARD PROBLEM

The study of awareness has been recently classified as a 'hard problem'; with the easy problems of awareness being ones that are nearly impossible to research[10]. The nature of mind has been called the oldest and most difficult problem facing human epistemology. It is only recently that a framework for posing the question of the nature of mind is reaching sufficient maturity[2,7,11]. Chalmers' initial premise that awareness is the fundamental principle from which to formulate a theory of mind[10] is a reasonable assumption; but going on the ask: 'what processes in the brain give rise to awareness?', creates the 'hard problem' because this manner of posing the question represents a category error for philosophy of mind. Historically whenever there has been a hard problem in science, it has turned out to be because the underlying principles have been poorly understood. Although it has been postulated that the mind/body is a naturally occurring form of conscious quantum computer; mind is more than brain or algorithm[11,12] and it is impossible to formulate a correct or sufficient theory of awareness from the point of view of AI, computer science or neuro-biology alone. Mind, to be adequately described, must be represented by a complete cosmology with mankind imbedded in it[2,7,8,9,13]. Currently about 93% of scientists believe the brain is sufficient to model the human mind.

4. PHILOSOPHY OF MIND - VITALISM / TELEOLOGY

The Continuous State Conscious Universe (CSCU) insists on reintroducing concepts like vitalism and teleology that have been historically disdained by science. Mechanistic models of the universe have had no place for these 'philosophical constructs' considered non-physical even by their proponents. In the CSCU they become physicalized and thus subject to falsification by empirical methods.

Teleology is the philosophy based on the supposition that the universe has design and purpose beyond the mechanics of a Newtonian or Big Bang universe driven acausally by a Darwinian type of natural evolution. Evolution obviously exists, but is guided by a teleological quantum of action inherent in the supra-locality of the CSCU. In perennial philosophies, teleology represents a basic argument for the existence of God, that the order and self-organization of the natural world are not accidental. If mind is fundamental to existence, an ultimate designer or teleological principle exhibiting a quantum of action must exist.

Modern teleologists like Hans Driesch or Henri Bergson proposed a principle of *vitalism* - that the processes of life result from, i.e. a self-determining fundamental rule not explicable by currently observed physiochemical laws. Bergson, proposed an *élan vital* or vital force[3] as the spontaneous energy of the evolutionary process and defined the mind as pure energy, responsible for all organic evolution; and denying the claim of science to explain the universe on purely mechanical principles. In ensuing sections this *vis vitae* is shown to be physical and discussed in terms of the noetic field[8,9].

5. PHILOSOPHY OF SCIENCE

In Newtonian clockwork mechanics consciousness was irrelevant or nonexistent. The advent of quantum mechanics introduced a troublesome observer involved in measurement. Although highly successful quantum mechanics has been deemed incomplete and does not describe biological systems suggesting that additional theory is required. The additional theory must describe the fundamental nature of time and include consciousness.

Traditionally five arrows of time[14,15] have been described:

- THERMODYNAMIC: The observation of temporal asymmetry in thermodynamic processes represents the most important arrow of time because it provides all of our phenomenological experience, and the existence of biological activity. Irreversible processes move toward a thermodynamic equilibrium of maximum entropy.
- EXPANSION OF THE UNIVERSE: Distinguishes between a past and future in the evolution of matter in the universe. In the CSCU this arrow

of time has no fundamental significance because it is an observational illusion proposed by an incorrect interpretation of astrophysical data[1].

- ELECTROMAGNETIC: It is generally observed, which leads to the belief that all electromagnetic waves propagate into the future only. However this is not necessarily true in interpretations of the transactional/absorber theory of radiation[16,17]; and represents a component of the illusion represented in Figure 4[1]. This is consistent with Maxwell's equations which are symmetrical in time.
- KAON DECAY: Nuclear reactions may occur in either direction with one exception occurring between elementary particles that are not part of ordinary matter - Neutral Kaons. There are three kinds of K meson but only the neutral Kaon exhibits a temporal asymmetry. Because neutral Kaon decay asymmetry is a spacetime property not associated with ordinary matter its description can be formalized into evidence of the supralocality of CSCU[18] by illustrating the variation of decay paths in terms of gravitational coupling to spacial and temporal nonlocal spacetime spin-exchange dynamics.
- PSYCHOLOGICAL: The subjective flow of time that reveals to our moment by moment experience that all actions flow from the present into what we define as the past. The contrast between the four arrows of time defined by physical laws and the subjective arrow of time has led to the belief that time is an illusion.

Currently physicists describe: 1. Physical time and 2. Psychological time separately, both of which are incompletely understood. A major premise of this work is that all five arrows of time are an illusion related to the phenomenology of mind and consciousness. But because mind in the CSCU is completely physical; all the arrows of time are likewise actually 'physical' and can be investigated with new experimental methods leading to discovery of the teleology of the noetic field. The unitary nonlocal noetic field couples classical dynamics and general relativistic effects in a complementarity through the pathways of neural dynamics. This is directly responsible for the perceived arrow of time because this mater-spacetime medium is what 'we' are made of and 'imbedded' in. The noetic timeless domain is the entry point of awareness that couples 'eternal time' through special relativistic dynamic transformations independent of classical gravitation to temporality.

6. THE STANDARD MODEL

While the nature of Minkowski/Riemann spacetime described by the Einstein 3-sphere is one of the most profoundly elegant descriptions of modern physics, and according to relativity theory "All methods lend themselves to the measurement of the curvature by experiments performed within the manifold, that is, with no need whatsoever to speak of an embedding of the manifold in a higher-dimensional space"[19]; the fact remains that relativity theory and quantum theory provide an incomplete description of the cosmology of the universe. While parallel transport, and the boundary of a boundary principle[20] being zero in the Bianchi identity as described by the Regge calculus seem to make the postulate of general relativity clear in terms of Miller's quote above; this is still an incomplete description, not describing conditions completely at the Planck scale or beyond the observed Hubble sphere.

The orientation of the zero boundary conditions through the 2nd law of thermodynamics provides the arrow of time for the Einstein 3-sphere causing our phenomenological reality to evanesce so to speak with apparent independence of the rest of the hyperdimensional structure beyond the Einstein description that contains it. Although this conceptual framework might seem fantastic in light of current thinking, lets try to clarify it. Firstly, the universe in the coming 'new gravity' does not reduce to a scalar Planck surface of compactified additional dimensions. The realm of our perceptual reality is the reduced space; our current conceptualization is thus reversed. An initial element of the new theory,[21] Spin Exchange Compactification shows that the so called higher dimensional complement is continuously compactified in time only, not in the higher hyperstructure where they remain fully extended. The purpose of these higher dimensions then, relative to our semi-collapsed reality is to provide a template for temporality to 'surf' on the face of 'eternity' through a continuous hyperdimensional translation of the extra dimensions or those excluded from our perception. We are 'flatlanders'[22] to the complete nature of the universe. This leads to an 'eternal continuous state universe'[1] with a broader continuum of temporal Hubble spheres co-moving within and without each other. In this harmonic transform there is rather than a reduction to a scalar Planck vacuum a continuous rotation into our collapsed phenomenological present of a transformation of the 1. classical, 2. quantum and 3. unitary as this present state is created annihilated and re-created out of the future-past. Our reality 'surfs' within the arrow of time by the 2nd law of thermodynamics and action of the noetic field. In the 'gaps' between the present-future/past and next present-future/past moment (like the spacing between frames on a film) the actions of the higher 'complete' 12 D hyperstructure takes place. Thus the constant c of the speed of light is inexorably connected to this spacetime/eternal transform and provides Earth with its proper time and us with mental clocks. The zero boundaries of the Einstein hyperstructure interface with the zero boundaries of the super-hyperstructure allowing them to interface and transform one within the other, with apparent chaotic stochasticity to our phenomenology and with perfect unity to the superstructure. The stochasticity and quantum uncertainty provides a perceptual phenomenological gulf between the two.

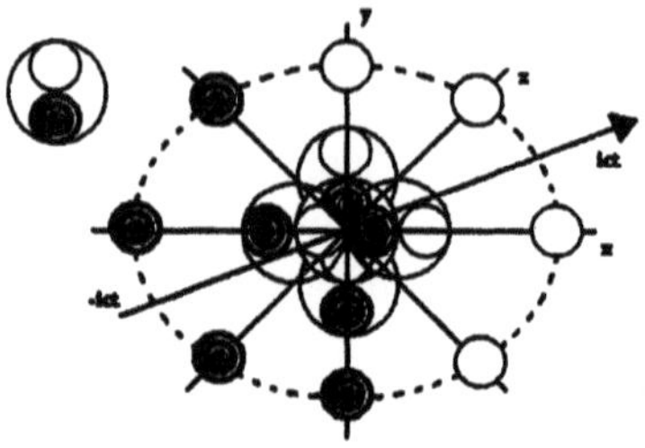

Figure 2. A 4-sphere of dimensions Δx, Δy, Δz, Δt partially separated at the center and its eight 3-sphere faces depicted as 'exploded off' of it. The exploded faces suppressed from the small figure at left. "The fourth-dimensional globe... is best indicated by circumscribing the numeral 8 inside a circle. So three circles are formed, the lower one representing the initial globe, the upper one representing empty space, and the greater circle circumscribing the whole. If it now be understood that the upper circle does not exist and the lower (small) circle is identical with the outer (large) circle..."[36]. In reality a single hypersphere does not exist independently. If we consider the upper white circle that Ouspensky calls empty space and non existent in terms of the Wheeler - Feynman - Cramer - Chu absorber theory of radiation in the light of the boundary of a boundary being zero, these so called empty spaces are conceived as orientations of the h-sphere in the unrealized future null direction.

The new CSCU model of the universe requires 3 hyperstructures like the Minkowski/Riemann Hubble 3-sphere to maintain itself. So while the profundity of the Minkowski 3 (4)-D structure of a boundary of a boundary canceling to zero through parallel transport[20] of the Bianci identity formulated by the Regge calculus as described by general relativity remains astonishing, the putative new superstructure must seem more fantastic by comparison since it challenges many of the well established principles of contemporary physics and cosmology. But it is logically consistent, and its explanatory power elegant. Its apparent antithesis with empirical evidence is one of interpretation only, not in antithesis to the data! The initial 'pin popping the balloon' of current thinking arises with the application of nonzero restmass photon anisotropy[21,23]. For example if photons couple to vacuum zero point energy during propagation then the perceived Hubble redshift does not signify a Doppler recessional velocity from an initial temporal singularity. It is a task beyond the conceptual presentation of this chapter to rigorously delineate the premises introduced; our purpose is merely to introduce the conceptual framework and leave the other issues to a future technical works.

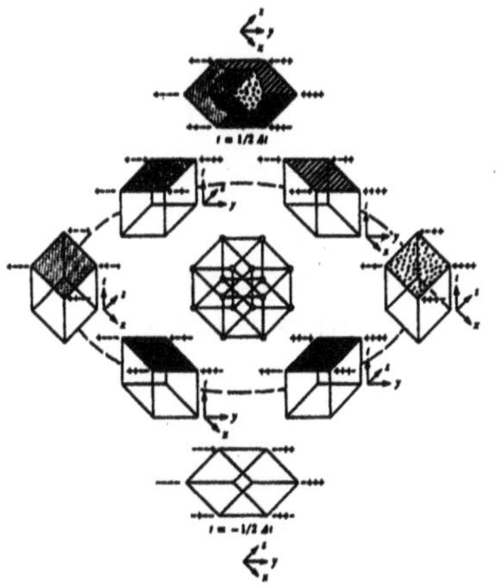

Figure 3. "A 4-cube of dimensions Δx, Δy, Δz, Δt and its eight 3-cube faces depicted as 'exploded off' of it. The content of the stress-energy tensor in the 3-cube at the upper right (shaded only at its 'top,' with diagonal stripes) measures the amount of momentum-energy that passes out of the 4-cube across a surface area $\Delta x\Delta y$, located at $z+{}^{\Delta z}/_{2}$, during the time Δt. To demand conservation of momentum and energy in all eight cubes, taken with due regard to sign, shall add to zero. This demand is fulfilled by representing the amount of momentum-energy in the given 3-cube (a vector or, in its dual representation, a 3-index tensor) as a sum of the 'moments of rotation' associated with all six 2-dimensional faces of that 3-cube. When we add up these contributions for all eight 3-cubes we have 6×8=48 2-dimensional faces to deal with, but their contributions all cancel out in pairs, as required. Thus the contribution of the striped face just considered is canceled by the contribution of the striped face at the 'top' or $\Delta x\Delta y\Delta z$ 3-cube. This is the payoff of the principle that the '2-dimensional boundary of a 3-dimensional boundary of a 4-cube' is zero. This is how spacetime grips mass so as to ensure and enforce the law of conservation of momentum and energy."

7. THE GALLILEAN, LORENTZ/POINCARE GROUPS OF TRANSFORMATIONS

Relative to our perception the universe is locally and globally an isotropic homogeneous 3-dimensional continuum; and other than the force of gravity there is no preferred direction. Simultaneity depends on the causality of the clockwork nature of events. This is the Gallilean group of transformations given by $x'=x+vt$, with the Euclidian metric $\Delta l^2=\Delta x^2+\Delta y^2+\Delta z^2$ where space and time are separate and absolute. The postulates of special relativity changed this - with simultaneity becoming dependent on the state of motion of the observer breaking down the absolute invariant separation of space and time producing a 4-dimensional continuum where physically equivalent observers become related by the well known group of Lorentz-Poincare transformations:

$$x' = \frac{x - \beta ct}{(1 - \beta^2)^{1/2}}, \quad y' = y, \quad ct' = \frac{ct - \beta x\, ct}{(1 - \beta^2)^{1/2}}, \quad z' = z \tag{7.1}$$

With $\beta = v/c$, c the speed of light and velocity v along the x axis and t and t' not identical. The symmetry group is defined by the point set

$$c^2 t^2 - x^2 - y^2 - z^2 = 0 \tag{7.2}$$

which by a Lorentz transformation maps into itself producing the light-cone of the invariant Minkowski 4-dimensional

$$\Delta s^2 = c^2 \Delta t^2 - \Delta x^2 - \Delta y^2 - \Delta z^2 \tag{7.3}$$

spacetime where every observer related by a Poincare' transformation has the same global spatial geometry extending to all of spacetime and physical events within the geometry have no effect on the geometry.

It is general relativity that allows matter to effect the geometry of spacetime; the Minkowski frames are local and

$$G_\mu^\nu = -8\pi k\, T_\mu^\nu \tag{7.4}$$

the curvature disallows its global extension. The geometry is governed by the 10 components of the Riemann curvature tensor relating to the distribution of matter and ensures that the lightcone structure has local Minkowski spacetime properties which is a solution to Einstein's equations.

8. GAUGE THEORY AND HIDDEN GAUGE INVARIANCE

It has long been considered that gauge invariance also called local phase invariance, must be associated with a massless gauge field. In QED this is the photon field A^μ. As generally known the gauge field allows the phase of all particles to be altered independently at widely separated spacetime points. The gauge field acts as a potential; and the relationship between the mass of the quantum and the range of the potential force leads to the conclusion that the gauge fields action over arbitrarily large spacetime distances must be considered an infinite range requiring gauge quanta to be massless[24]. Atchison and Hey go on to show how masslesness for gauge quanta is rigorously linked to gauge invariance. First by showing that the electromagnetic potential satisfies the Maxwell equation

$$\Box A^{\nu}-\partial^{\nu}(\partial_{\mu}A^{\mu})=j^{\nu} \qquad (8.1)$$

where $\Box\equiv\partial_{\mu}\partial^{\mu}=\partial^2/\partial t^2-\nabla^2$ is the D'Alembertan operator for spin-0 bosons; and is shown to be invariant under the gauge transformation

$$A^{\mu}\rightarrow A^{\prime\mu}=A^{\mu}-\partial^{\mu}\chi \qquad (8.2)$$

Finally, if A^{μ} represented a massive field, the resultant

$$(\Box+M^2)A^{\nu}-\partial^{\nu}(\partial_{\mu}A^{\mu})=j^{\nu} \qquad (8.3)$$

is not invariant in terms of (8.2) because the introduction of the mass term M^2A^{ν} breaks the gauge invariance[24].

In spite of this solid argument for non invariance by the introduction of mass terms it is possible for gauge quanta to acquire mass and not lose gauge invariance if the gauge invariance is hidden and the mass has a special relationship with its couplings[24]. Anderson showed that gauge bosons can only acquire mass if the symmetry is hidden.

The path to hidden gauge invariance is long and tedious and remains controversial. In depth discussion can be found in the literature; it is our purpose here to assume hidden gauge invariance that doesn't violate standard gauge theory and discuss the interesting consequences for cosmology and reduction of the wave function in ensuing sections.

9. THE COSMOLOGY OF A CONTINUOUS STATE CONSCIOUS UNIVERSE (CSCU)

It is assumed that the fundamental nature of time cannot be understood from within the current standard model of a Big Bang Universe and the conceptual foundation of physical theory needs to be "Godelized" beyond it. Quantum Theory is the cosmology of matter and radiation that superceded Newtonian classical mechanics. Matter is imbedded in spacetime and it is also assumed that time asymmetry is more fundamental than quantum theory[25] and that time emerges from a more fundamental unitary domain[26].

The two main pillars of the Big Bang are based on the properties of light - that of the observed Hubble redshift and the Cosmic Microwave Background Radiation (CMBR).

The CSCU is based on extended electromagnetic theory (EM) where the photon has additional longitudinal degrees of freedom and mass anisotropy[21,27]. This means that the observed Hubble redshift is not Doppler but arises from coupling of the photon to vacuum dynamics during propagation. This is the Vigier theory of 'tired-light'. Likewise the CMBR in CSCU did not originate in a primeval explosion but occurs continuously as a black body radiation in spacetime gravitational cavity dynamics[18,28]. The details of which are beyond the scope of this chapter and are discussed elsewhere[1].

The higher 'extra' dimensions were not compactified in an initial big bang singularity, but are continuously compactified and recompactified in a spin exchange process relative to our perceived reality in a process which creates the physical nature of the 'illusionary' arrow of time. Further properties of the CSCU are reminiscent of Kant's antinomy of spacetime which he proposed as a solution to the argument between Newton and Liebniz as to whether the universe was open or closed. The observed Hubble radius Einstein 3-sphere of our perceptual reality is closed and finite temporally; but is open and infinite atemporally. Thus the CSCU is like an 11-(12)-D hyper Klein bottle with an infinite number of Hubble type spheres nested within it each of which might have a variance of the laws of physics[23]. This is consistent with extended Everett/ Bohm modeling and the Wheeler/Feynamn/Cramer/Chu transactional many worlds theories[16,17,29].

The 12 Dimensional CSCU is comprised of three 3 (4) sets of Einstein/Minkowski/Reiman spacetime packages consisting of a complement of three temporal dimensions. 8-(9) of the 12 dimensions are nonlocal or 'hidden' at any momentary slice in the standing wave of spacetime. This is represented conceptually in Figure 4. Here spacetime does not reduce to Planck scale compactified dimensions as in the current reductionist standard model. In CSCU cosmology our Minkowski/Riemann space is a subspace imbedded in the higher complete twelve dimensional space. The continuous harmonic translation of the symmetry of these canceled (boundary of a boundary equals zero) boundary conditions (relative to our perception) allows nonlocal contact of aspects of the Minkowski/Riemann 3-(4)-D hyperstructure within the 'perfect' 11-(12)-D hyperstructure to be invisible to our phenomenology (because boundary of boundary equals zero) and to contain and sustain it.

DEEPER ASPECTS BEYOND THE STANDARD MODEL INHERENT IN THE DUALISM OF QUANTUM THEORY

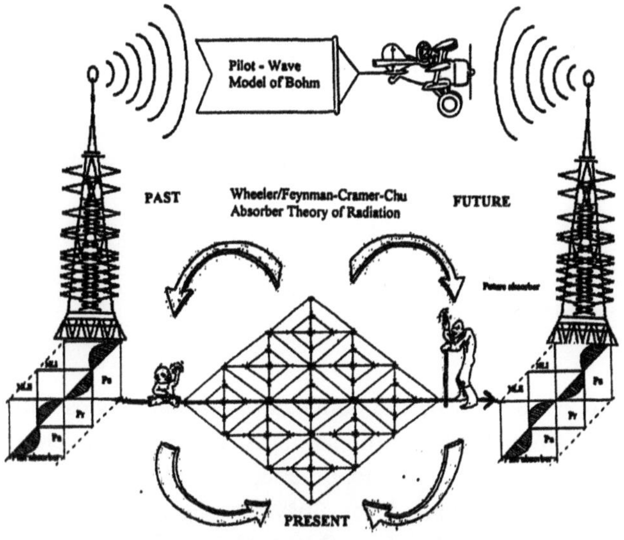

Figure 4. The quantum field of information may be compared to a ship on automatic pilot guided by radio waves from the future and the past. In addition to the local/nonlocal standard model of quantum theory; a deeper nonlocal/unitary ontological aspect suggests that a quantum dualism is required for a complete quantum theory. The ontological domain is described by a combination of the Everett/Bohm Pilot wave non-collapse version and the Wheeler/Feynamn Cramer/Chu transactional-absorber theory of radiation. In this model the present collapses from a higher dimensional hyperstructure continuously recreated out of the energy of the future-past. The pilot-wave quantum potential has a relationship to causality. Gravitational tidal effects regulate aspects of quantum state evolution U and state vector reduction R in both collapse and non-collapse domains of the quantum dualism. This cosmological spacetime structure is an important aspect of the conscious universe; and is required because the standard model of quantum theory is not sufficient to describe the mind. It is through this deeper structure that the Noetic field enters the body and there are enough degrees of freedom for reason to operate. In the figure much of the hyperstructure is suppressed to give a 2D view. The three 3-(4)-D spacetime packages are spread out from left to right across the figure. The central unit labeled the present in terms of our 'reality' consists of a superposition of all three in a dynamic hyperdimensional 'standing wave'.

Time is considered a structurally incomplete aspect of our perceptual phenomenological reality. While geometrodynamics is sufficient to describe coordinate free particle paths in terms of the Bianchi identity and parallel transport around a closed curve; as is well known general relativity does not provide a complete description of gravity. As will be described the additional 3(4)-D packages are necessitated by the introduction of sub-millimeter gravity and the cosmology of consciousness.

10. THEORY OF MIND

Noetic field theory[2,7,8,9] suggests that the mind $|\psi_M>$ is a state with quantifiable superimposed properties. This phenomenology/noumenon of consciousness is composed of three integrated base states: Elemental intelligence $|\psi_e>|$, Cosmological ordering principle $|\psi_c>$, and the brain defined as an apparatus $|B|\psi_M>$.

The base states of mind interact at the quantum and nonlocal levels as described conceptually in equation 10.1 or as the summation represented in

$$|\psi_M>=|\psi_e>+|\psi_c>+|B|\psi_b> \tag{10.1}$$

$$|\psi_M>=\sum_i^{Z^\alpha \bar{Z}_\alpha} N_i|\psi_i> \tag{10.2}$$

equation 10.2 where N is the superimposed base states summed over a prespace twistor singularity (Nonlinear aspects appear on expansion).

The domain for sensory phenomena, mentation, and all noumenal conscious processing is called the 'psychosphere'. The structure of the psychosphere is the complex bound containing the totality of individual consciousness and the extent of its influence. This means both locally in the brain and body fields, and nonlocally in prespace. The psychosphere includes a higher dimensional hypersphere or tesseract for nonlocal interactions. This is the domain from where coherent Bose psychons condense. Once it is realized that there is more to mind than brain or algorithm a comprehensive theory of consciousness can be formulated in terms of a continuous state conscious universe (CSCU). The mind is comprised of three base states: 1. Classical and quantum brain dynamics, 2. A nonlocal elemental intelligence that serves as an atemporal boundary condition of individuality and 3. The interaction of the nonlocal noetic unified field that integrates the duality[2,7,8,9,13]. Recent studies of extended electromagnetic theory (EEM) claim that this work could lead to the 'completion' of quantum theory and that it will be a deeper understanding of small scale gravitational effects that will be instrumental in this breakthrough. The correspondences of this new theory are presented and a description given of how this relates to the parameters that comprise times arrow and how it is related to our psychological perception of time.

The three-fold nature of noetic field theory is best symbolized by a Penrose triangle. See Figure 5A. The three base states are not separable in the 'standing wave' of reality space. In figure 5B, a modification of the form of Penrose's three worlds to be compatible with noetic theory, it can be envisioned how the three NFT base states cycle continuously into our spacetime from the twelve dimensional unitary domain that our reality is a subspace of. The reader should understand figures 5A-B in terms of figure 4 to better conceptualize the CSCU.

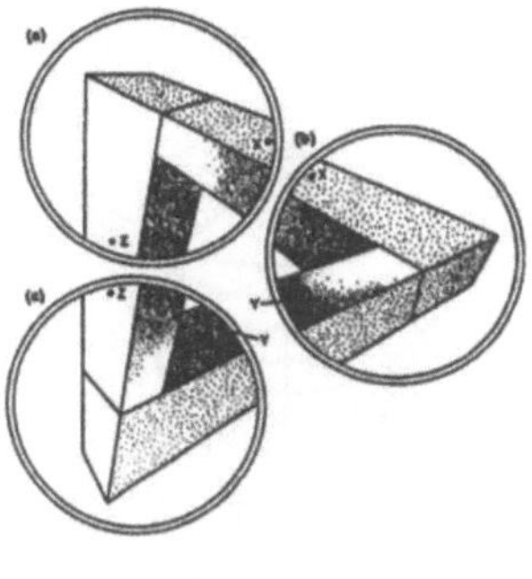

Figure 5A. The Penrose Triangle Impossible in three dimensions.

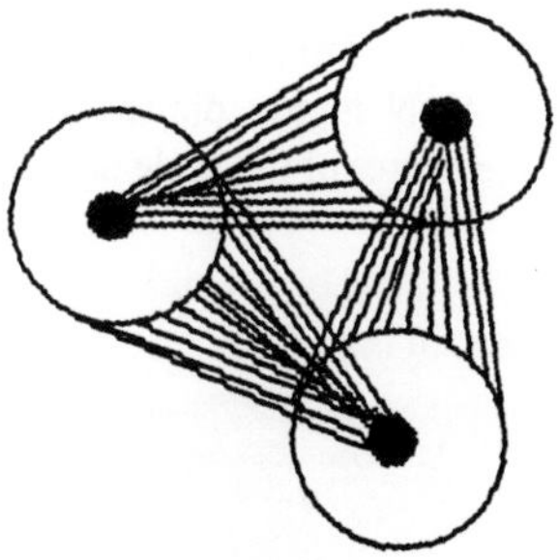

Figure 5B. The three worlds of NFT (Modified from Penrose 3 worlds, 1994).

The CSCU is considered scale invariant[23]; the teleology of which has self-organized the embedding of our consciousness into a central perceptual position of the Hubble sphere as illustrated in Figure 6. In Figure 7 a Holophote (lighthouse) is used to illustrate the *elan Vital*[3] or life-force and how it enters spacetime and every atom of our being to self-organize life and the 'light of our mind'. Using the metaphor of the holophote, the noetic 'source' enters spacetime from the unitary domain through parallel transport in the spinor structure[30] of the 'lens' and reflectors in the 12D hyperstrtucture. The rotation of the reflectors is symbolic of the dipole oscilation of the oscillating EM field. Interestingly the Frohlich frequencies in the brain have been found to be the same as those in the CMBR. Supralocality is outside the circle of the arrows; and the appearance is inward towards the center keeping with the CSCU supposition that points in our reality are in a subspace of supralocality.

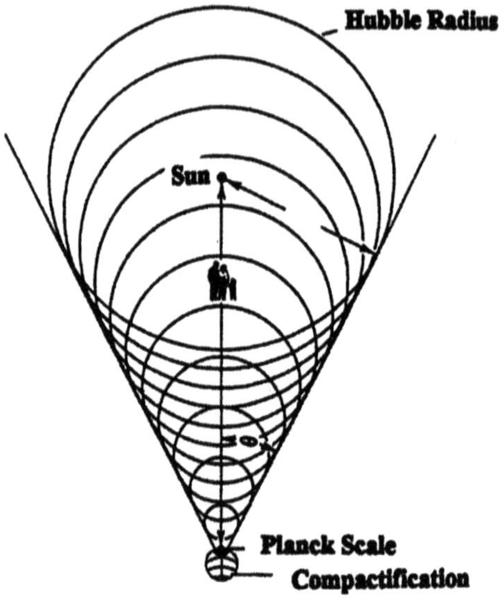

Figure 6. Current SM view of Universe from Hubble radius H_R reduced to a Planck scale spacial singularity M_{pl}^{-1} imbedded in N higher dimensions fixed in a Big Bang temporal singularity with the Man-scale at the center of the scale invariance.

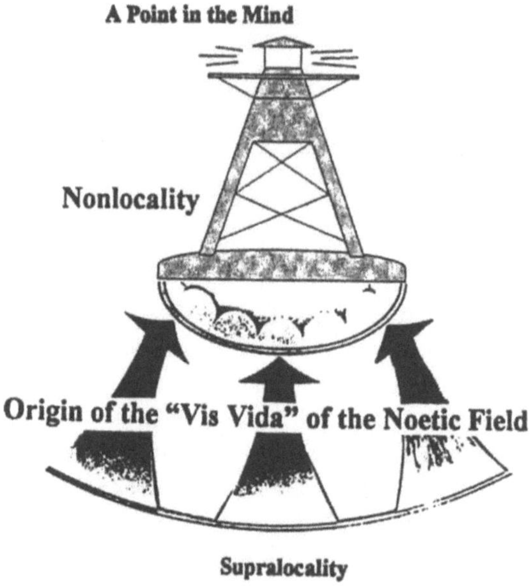

Figure 7. Holophote (light house effect) transition of unitarity by high frequency oscillating noetic field and 'elan vital' of consciousness. The physical structure projects the arrow of time from unitarity –NL-III through NL-1 & NL-II symbolized by the legs of the tower.

11. THE ORIGIN OF TIME IN THE CSCU

Now that the model of consciousness has been conceptually illustrated, we are able to show how the 'flux tube' of time is generated into our level of 'virtual phenomenological' reality. We have supposed that our reality is comprised of a standing wave of hyperdimensional spinor dynamics acting in concert in three Minkowski/Riemann 3 (4)-D spacetime packages that together form the 12D supralocality of the complete CSCU. Because by parallel transport all the boundary conditions sum to zero at each present moment according to the Wheeler-Feynman/Cramer-Chu absorber/transactional model; we are comprised of the 'matter' projected from this fundamental space, and our consciousness is likewise coupled to and imbedded in this same material by a "philosophical tension"[31] caused by the flow of awareness in conjunction with thermodynamic evolution of matter and the teleology of the noetic field that mediates it. This structure is depicted in Figure 8. Two 3 (4)-D packages are suppressed.

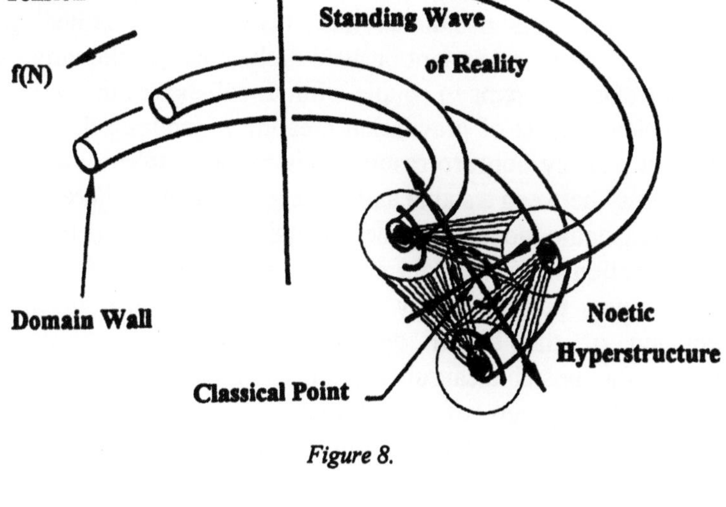

Figure 8.

12. ISSUES SURROUNDING STATE-VECTOR REDUCTION

Generally any quantum state may be reduced by the phenomenology of measurement according to the orthodoxy of Schrodinger evolution. This has been called the stable causality condition. There is no quarrel with this firmly established principle as it relates to a typical measurement where a particle or wave impinges a 'detector'. However spatially or less commonly temporally separated states of freely propagating test particles may become correlated through a weak nonlocal causal action of the quantum potential. Generally according to the locality principle there is no direct causal connection between points with space-like separation except for certain EPR pairs correlated nonlocally with the same unique spacetime package induced within the simultaneity of the initial pair production as in dual photon emission from a mercury atom.

But the case to be discussed here is a mechanism of geodesic deviation whereby initially separated systems without any initial EPR correlation may become superposed through the gravitational action of nonzero restmass photon anisotropy and longitudinal Extended Electromagnetic interactions (E-EM)[9]. Shimony[32] states "it is impossible to capitalize upon the entanglement of the quantum state of a two-particle system for the purpose of sending a message to an observer of one of the particles by performing an operation upon the other particle. This is a theorem which has been proven in great generality by Eberhar, Ghirardi et al., and Page". It is the proposal here that this premise is true only by the postulates of quantum orthodoxy for more classical type states and not quantum ontology where the uncertainty principle may be clearly violated under certain 'non-collapse' conditions.

In this chapter it is postulated that there is an additional group of transformations with a variant causal relationship allowing spatially separated correlated systems to couple without collapse of the wave function through the tidal effects of gravitation creating a geodesic deviation in the normally null path by nonzero restmass photon anisotropy. This requires a complementarity between a stable and unstable causality, where the null path of the standard stable models may be maintained through the connectedness of the conservation of momentum and the validity of the CPT theorem. Or this causality may be violated by nonlocal gravitational interaction in the transformation of the higher dimensional hyperstructure through the introduction of an unstable causality and disconnectedness. This leads to the need for another group of transformations beyond the Galilean, & Lorentz-Poincare' groups; the description of which will lead to particle dynamics beyond the standard models that includes transformations of the flow of consciousness.

Roger Penrose 'On gravity's role in quantum state reduction'[33] gives an account of the status of the theoretical relationship of the role of a putative gravitational effect related to state vector reduction. Penrose is forced to conclude that the standard Copenhagen interpretation of quantum theory (QT) is incapable of defining any role for gravity. Penrose examines the quantum superposition of two different stationary mass distributions not in a specific spacetime but in a superposition of two slightly differing spacetimes. In all models of QT, phenomenological or ontological, the quantum state is considered to evolve according to forms of the Schrödinger wave equation.

$$ih\frac{\partial\psi}{\partial t} = -\frac{h^2}{2m}\Delta^2 + V\psi \qquad (12.1)$$

Issues that some physicists consider as needing further elaboration center around the nature of quantum objects, state reduction as it relates to the measurement problem, and whether QT needs to be extended to include non-collapse versions of the Everett/Bohm type. "...the resolution of the problem of state-vector reduction in quantum - mechanics should be one of the major tasks of a (successful) quantum gravity theory. This seems to be very much a minority view...It seems that many physicists do not even accept that there *is* a problem of state-vector reduction!"[33]. The noetic group of transformations requires an ontological non-collapse extension of quantum theory.

In Everett's well known paper[29] from which the 'many-worlds' interpretation of QT originated; Everett makes the assumption that the state of a system $|\psi\rangle$, evolves only according to the time-dependent Schrodinger wave equation:

$$ih\,\partial/\partial t\,|\psi\rangle = H\,|\psi\rangle \qquad (12.2)$$

13. NON ZERO RESTMASS PHOTON ANISOTROPY AND PHOTON STRUCTURE

The hidden gauge invariance for the photon occurs as an anisotropic restmass of $\sim 10^{-65}g$ that periodically couples to the vacuum zero point fluctuation[27]. Although there has been discussion of non zero photon restmass for many years it has typically been ignored because as stated above it was considered in violation of Gauge theory and also to be ad hoc. Recently it has been shown[21] that the photon has mass in terms of considerations from the weak field limit and quantum gravity, both of which can be derived from first principles.

According to Einstein rest energy or mass is the result of external or internal structural motion of a particle. Unlike Fermi matter that maintains a well developed internal kinetic structure through atomic movements whether at rest or in motion, photons cannot be brought to rest without immediate annihilation and dissipation of their energy because its symmetry cannot be maintained at rest. For propagating photons the internal transformation undergoes anisotropic oscillation. The restmass fluctuates harmonically from zero to > 0 as a consequence of the complementarity of its wave particle duality; which signifies according to $E = MC^2$ a change in energy from inward reflection during the instance of mass and interaction with the vacuum to outward displacement through space during zero mass moment. In addition to the wave - particle duality photons also have complementarity between locality and nonlocality in there propagation. They exist both in spacetime and independent of it; therefore the timelessness of the photon is maintained during the masslessness of its wave propagation at the constant speed c, allowing the temporal particle moment coupling to the vacuum zero point oscillation to not interfere. Fluctuation in mass-energy is not mysterious as it is generally known that inertial and gravitational masses are an aspect of this movement[34].

"In other words, the transformation of "matter" into "energy" is just a change from one form of movement (inwardly, reflecting to-and-fro) into another form (e.g., outward displacement through space.) The possibility for objects of zero rest mass exists provided that they are moving at light speed. For if rest mass is "inner" movement, taking place even when an object is visibly at rest, it follows that something without "rest mass" has no such inner movement, and that *all* its movement is outward, in the sense that it is involved in displacement through space. Light does not have the possibility of being "at rest" because it does not possess any such inner movements"[34].

14. TIDAL EFFECTS OF NEWTONIAN GRAVITY

Based on the critical assumption of non zero restmass photon anisotropy, it becomes straight forward (conceptually) to apply the tidal effects of Newtonian gravity to produce a superposition of two spatially separated coordinate systems without collapse of the state vector. This means that there is a dual aspect to state vector reduction - one that applies classically in terms of the standard model, and another in terms of Everett/Bohm non collapse models by the action of the (B3) longitudinal field. The topology changes with mass. The tidal curvature change allows two spatially separated geodesics to converge into one spacetime and under certain phase conditions to superpose without collapse; which is a nonlocal effect of vacuum zero point fluctuations.

The evolution of the quantum state described by the Schrodinger wave equation (12.1; 12.2) and its reduction is considered to be the core of quantum theory (QT). It is suggested that the Copenhagen interpretation of QT provides an incomplete description of the physical world; and that the application of non zero restmass photon anisotropy leads directly to a dual description of state vector reduction in terms of both the standard model and non collapse versions of QT which are found not to be in violation of hidden guage invariance in Guage Theories as typically considered. In the Poincare' group the simplest realization is the geometrical interpretation of the free particle trajectory as a world line. It is suggested that the application of non zero restmass photon anisotropy leads directly to a geodesic deviation in apparent violation of the standard Copenhagen interpretation, but not for ontological noncollapse versions. This requiures a desription of strong and weak causal factors that relate to separated systems that are correlated and uncorrelated and suggests that there is another group of transformations beyond the Poincare' group needed to incorporate the higher dimensionality beyond the Minkowski/Reimannian metric. This has exciting implications for the completion of QT that will lead to a deeper understanding of the requirements for developing Planck scale Geometrodynamics.

15. THE COUPLING OF THE NOETIC FIELD

It is generally known that the standard models of quantum theory and cosmology do not include consciousness or give an adequate description of the nature of time; suggesting that the elucidation of these ideas must come from extended theoretical insights. Human perception indicates a flow of time - from past, to present to future in accordance with the second law of thermodynamics by appropriate changes in entropy of the system observed as it undergoes evolution. Thus temporal order seems related to entropic order; and these dynamics constitute how we perceive 'action' or translation in this particular dimension, the dimension of time.

In terms of the vertical or up/down dimension, 'action' or direction is determined in our awareness by the nature of the gravitational force. In the absence of gravitation one loses all 'sense' of this direction or dimension. Can this relationship of the force of gravitation to direction be equated to a similar action for the perception of time? Is there a force or inherent 'action' in the cosmology of mind that couples awareness to the entropic activity observed in the perceived external reality that we equate with the flow of time? Do we 'ingest' time parameters through an 'axis mundi' into our psyche during transit through the cosmological sea as a whale ingests plankton while swimming in the ocean? If it is true, that the entire perception of time is a creation of a normative human ontology through an innate or habituated philosophic tension[31] that drives the orientation of our mind, a decoupling from the flux of this noetic field would allow a reorientation of our 'psychosphere'[2,8,9] and the possibility to re-tune the perception of our psyche to additional or alternative parameters of entropic 'action', time or atemporality. This is a key point in understanding the nature of time and will be described in more detail.

16. THE NOETIC GROUP OF TRANSFORMATIONS

This paper postulates additional transformations here called the 'noetic group' with an another causal relationship distinct from the strong causality of the standard model allowing spatially separated systems to exchange information without orthodox collapse of the wave function. This occurs through a nonlocal gravitational coupling of (B3) field effects which produce a geodesic deviation mediated by intentionality. The dynamics of particles and fields are described by various groups of transformations; the Galilean group describes Newtonian mechanics, and the Lorentz transformations describe modern theory. This action is outside the current limits described by the Galilean, Lorentz and Poincare groups of transformations. This additional noetic transformation of a normally null path is allowed in extended electromagnetic theory by nonzero restmass photon anisotropy[21] without violating gauge theory[24]. The correspondences in physical theory, for example the reduction of quantum mechanics to classical mechanics or the recovery of thermodynamics from its successor statistical mechanics will also apply in relation to the mind and the nature of time. A correspondence between a complementary stable and unstable causality is shown to reduce to the null path of the standard model.

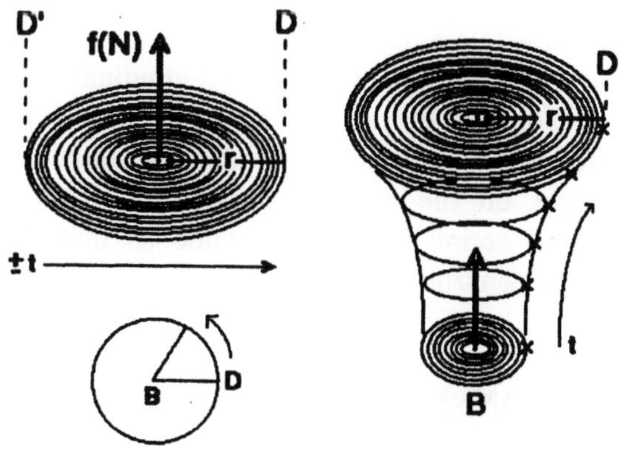

Temporal/Atemporal Duality in the Wavefunction of Existence

Figure 9. A conceptualization of the complementary structure of the wavefunction of an entity. F(N) is the noetic quantum of action supplying the 'philosophical tension'[31] that couples consciousness to the 'physicality' the entity is imbedded in. B is the point of entry or birth of the conscious entity into local spacetime; and D is death or exit from the domain. R is the psychosphere radius of the entity symbolizing the coherence length and content of the mind. In the smaller reference circle it can be seen that r evolves with age from B to D; although the radial velocity is equal, the maturer entity maps out more area in the same time. This explains why timeperception seems longer for a youth and accelerates for an adult. In an eternal sense the entity would 'oscillate' around some average position of its wave function. An entity may acquire an infinite amount of information; death does not come with a limit of information but because stressors have a tendency to translate the coupling loci (represented by the X's in the exploded figure) until the associated 'de Broglie wave of consciousness' reaches the limit of the domain wall of the 'wave function' and the 'spirit' tunnels out of the Minkowski domain of our reality.

This new group of noetic transformations is a complementarity of NL-I & NL-II dynamics, which is an element missing in the typical EPR experiments done with classical - either temporal or spatial measuring devices because of the action of the uncertainty principle in the reality of our subspace. The mind is not a classical device, but a whole cosmology encompassing both NL dynamics simultaneously: The standard classical either or interactions acting in brain phenomenology and an ontological non-collapse complement acting in the noetic mind. In CSCU cosmology an additional causal principle is involved in the noetic group of transformations. Its action occurs through the nonlocal teleology and action of human intentionality[9]. These CSCU transformations act in an 'event-like' manner when the correspondence is relative to locality and are energyless interactions as described by extended electromagnetic theory (E-EM) and gravitational interactions[9] No particle mediates the energyless interactions which exchange information by Bose-Einstein superposition.

The action of intentionality modulates the spin alignment angular momentum of the noetic field. This constitutes a gravitational action on infinitesimally separated geodesics by the longitudinal action of the (B3) field that does not collapse the wave function. This is an example of the non-computable actions of the mind/body interface where intentionality acts as a phase regulator on the holoscape manifold of Pribrams dendritic microprocess[9,13]. This action is also the same as that described by the psychon action on the dendron by Eccles.

17. FINAL CONSIDERATIONS

We have seen that the application of non zero photon restmass provides a simple method to describe a quantum dualism that includes both collapse and non collapse models of quantum state reduction. This has broad implications for completing QT, setting the stage for a deeper understanding of Geometrodynamics at the Planck scale and an understanding of noncomputability as it might relate to advances in quantum computing and applications to understanding the nature of consciousness and time.

Both general relativity and quantum theory are known to be incomplete. Twistor and superstring theories are currently considered the most promising candidates for their nonlocal integration; but both lack a 'Rosetta stone' for delineating the unique topological package of higher dimensional hyperstructure required to complete the task. Both eastern and western theologies claim that gravitation is caused by the movement of spirit - spirit, ki, chi, or prana is not immaterial but Bose or photon based in Noetic theory. The spin exchange model of quantum gravity[21] incorporating the expanded Wheeler- Feynman absorber theory of radiation[17,35] putatively describes gravitons as superposed moments of confined nonlocal photons mediated by unitary dynamics.

The cosmological constant is the coupling constant between both domains; the zero averaged fluctuation of the gravitational potential localizes and delocalizes the flow of conscious energy. The lower limit for the quantization of mind is a Planck scale hypercavity where the gravitational potential may remain balanced when at rest. Gravitational mass dependency is not required by conscious entities for state evolution as in the Hameroff-Penrose Orch-Or model because spacetime curvature provides boundary conditions gating the energy flow of Bose psychons. No gravitational work is required, mental activity is frictionless as in the (B3) field at this level, but not at the higher organic species level. Radiation or light pressure is sufficient to modulate the boundary conditions. All levels of scale are proportional to the elemental Planck unit through the law of energy quantization. Degenerate energy from infinite density singularities not being applicable to consciousness. Thus one may whimsically query 'How many Einstein's (moles of photons) does it take to turn on a light bulb? The bulb being a 'byte' of Planck bits pertinent to the conscious scale of the entity. Thus entity Z with a 10^N Planck byte raster of consciousness, has a 10^N byte psychosphere and resolves factors of 10^N bytes of external and internal mental events. Thought being dynamic moments of local quantization and summation of conscious energy. This is the cosmological root of consciousness.

18. SUMMARY

Noetic theory gives a full explanation of the origin of time, a comprehensive description of consciousness and experimental tests for isolating the noetic field have been prepared. 1. Existance is in terms of a continuous state conscious universe (No Big Bang) 2. This means that there is a nonlocal teleological principle that has causal action on both cosmology and living systems. 3. This further entails a *vis vitae* or *elan vital* - vital force that in terms of noetic theory is PHYSICAL and therefore falsifiable. 4. This vital force is the noetic field, a quantum of action which is associated with the unified field and chi, prana, etc in common nomenclature. 5. Because the noetic field is associated with the unified field it is also coupled to small scale gravitation which introduces the causality of consciousness both in terms of the universal teleological principle, and in terms of Man's consciousness and agency. 6. This model besides new cosmology requires extended quantum, electromagnetic and gravitational theory. The foundation for most of which is already known. 7. All of physics is based on measurement and duration using yardsticks composed of the same material that we are made of and therefore steeped in our consciousness. Subjective and phenomenological Reality is therefore an illusion (as we are coupled to it in our ordinary wakeful state). Whether this seems hard to swallow or not at present, this is the most fundamental requirement to embrace if we are to get bet beyond or 'Godelize' sufficiently to ever understand both the nature of time and the nature of consciousness. 8. Therefore we assume that ALL of physics is also falsified up to this 'classical' limit (same as Newton was falsified). 9. Therefore time is an illusion also. All the 5 currently described arrows of time are not psychological per se but rather functions of the consciousness or the teleology of the conscious universe; but because in the continuous state conscious universe of noetic field theory consciousness is physical, all the parameters of the arrows of time are likewise physical.

10. Opposed to reductionism of the standard models, the compactified dimensions were not laid down in the big bang to exist at a Planck scale 'below' us. In contrast we are a subspace of a higher 11-(12)-D universe in continuous transformation (continuous state). This is very important because this sets the foundation for how the physical arrows of time and how the noetic *elan vital* is 'projected' into our subspace 3-(4)-D domain. This is what is to be tested in the experimental protocol: The entry point of the noetic field and the structure it gives to spacetime and consciousness.

REFERENCES

1. Amoroso R.L., 2000, The Continuous State Conscious Universe, *Vigier III Symposium*, UC Berkeley.
2. Amoroso R.L., 1997, Consciousness, a radical definition, *Noetic Journal*, Vol.1 **1** 19-27.
3. Bergson H., 1911, *Creative Evolution.*
4. Kafatos M. and Nadeau R., 1990, *The Conscious Universe*, Springer-Verlag, New York.
5. Prigogine I., 1996, *The End of Certainty*, The Free Press, New York.
6. Einstein A., 1950, *The Meaning of Relativity*, Princeton Univ. Press.
7. Amoroso R.L., 1997, A Brief Introduction to Noetic Field Theory: The Quantization of Mind, *Brain and Consciousness*, Rakic, Kostopoulos, Rakovic, and Koruga (eds.), ECPD: Belgrade 297-302.
8. Amoroso R.L., 1999, An Introduction to Noetic Field Theory: The Quantization of Mind, *Noetic Journal*, Vol. 2 **1** 28-37.
9. Amoroso R.L., 1999, The Physical Parameters of Intensionality, *Address at 2nd GMU workshop on Fundamental Principles of Philosophy*, to be published.
10. Chalmers D., 1996, *The Conscious Mind*, Oxford University Press.
11. Amoroso R.L., 1997, Engineering a conscious quantum computer, *Mind Versus Computer*, Gams et al. (eds.), IOS Press.
12. Amoroso R.L., 1995, The extracellular containment of natural intelligence, *Informatica*, **19** 585-590.
13. Amoroso R.L. and Martin B., 1995, Modeling the Heisenberg matrix: Quantum coherence and thought at the holoscape manifold and deeper complementarity, *Scale in Conscious Experience*, King and Pribram (eds.), Mahwah: Lawrence Earlbaum.
14. Morris R., 1984, *Time's Arrows*, Simon and Schuster, New York.
15. Davies P., 1974, *Physics of Time Asymmetry*, Univ of California Press, Berkeley.
16. Wheeler J.A. and Feynman R.P., 1945, *Rev. of Modern Physics*, **17** 157.
17. Cramer J.G., 1986, Transactional interpretation of quantum mechanics, *Rev. of Modern Physics*, Vol. **58** **3** 647-687.
18. Amoroso R.L. and Vigier J-P., 2000, *The Origin of the Cosmic Microwave Background*, preprint.
19. Miller W.A., 1985, The geometrodynamic content of the Regge equations as illuminated by the boundary of a boundary principle, *Between Quantum and Cosmos*, Zurek, A.van der Merwe, and Miller (eds.), Princeton Univ. Press.
20. Cartan E., 1979, *Albert Einstein Letters on Absolute Parallelism 1929-1932*, Debever (ed.), Princeton Univ. Press.

21. Amoroso R.L., Kafatos M., and Ecimovic P., 1998, The origin of cosmological redshift in spin exchange vacuum compactification and nonzero rest mass photon anisotropy, *Causality and Locality in Modern Physics*, Hunter and Jeffers (eds.), Kluwer Academic, Dordrecht.
22. Abbot, E.A., 1983, *Flatland*, Harper Perennial, New York.
23. Kafatos M., Roy S., and Amoroso R., 2000, Scaling in cosmology and the arrow of time, *Studies on the Structure of Time: From Physics to Psych(patho)logy*, Buccheri and Saniga (eds.), Kluwer, Dordrecht.
24. Atchison I. and Hey A., 1982, *Guage Theories in Particle Physics*, Adam Hilger, Bristol.
25. Wooters W.K., 1994, Is time asymmetry logically prior to quantum mechanics?, *Physical Origins of Time Asymmetry*, Halliwell, Perez-Mercader and Zurek (eds.), Cambridge Univ Press.
26. Barbour J.B., 1994, The emergence of time and its arrow from timelessness, *Physical Origins of Time Asymmetry*, Halliwell, Perez-Mercader, and Zurek (eds.), Cambridge Univ Press.
27. Vigier J.P., 1992, Status of the Einstein, de Broglie theory of light, *Preoceedings of the ISQM workshop on quantum mechanics*, Tokyo.
28. Amoroso R.L.,1997, *The Origin of Cosmic Microwave Background Radiation in the Intrinsic Fluctuation of Cavity Quantum Electrodynamics*, preprint.
28. Everett H., 1957, *Rev. Mod. Phys.*, **29** 454.
30. Cartan E., 1966, *The Theory of Spinors*, New York, Dover Publications.
31. Draganescu M., 1991, *The Philosophical tension and the cosmic feeling*, The Romanian Academy, Bucharest.
32. Shimony A., 1985, Events and processes in the quantum world, Penrose and Isham (eds.), *Quantum Concepts in Space and Time*, Oxford Univ. Press.
33. Penrose R., 1996, On gravity's role in quantum state reduction. *Gen. Rel. and Grav.* 28, 5, 581-600.
34. Bohm D., 1965, *The Special Theory of Relativity*, New York, W.A. Benjamin.
35. Chu S.Y., 1993, *Physical Review Letters* **71** 2847.
36. Ouspensky P.D., 1970, *Tertium Organum*, Vintage Books, New York.
37. Abolhasani M. and Golshani M., 1999, Born's Principle, action-reaction problem and arrow of time, *Found Phys Let*, 12:3, 299-306.
38. Amoroso R.L., 1996, The production of Frohlich and Bose-Einstein coherent states in in vitro paracrystaline oligomers using laser phase control interferometry, *Bioelectrochemistry and Bioenergetics*, **40** 39-42.
39. Barut A.O., 1980, *Electrodynamics and Classical Theory of Fields and Particles*, Dover Publications, New York.
40. Ben-Dov Y., 1990, No-collapse versions of quantum mechanics, *Quantum Theory Without Reduction*, Cini and Levy-Leblond (eds.), Adam Hilger, New York.
41. Costa de Beauregard O., 1999, Lawlike reversibility, factlike irreversibility, The symmetry of time in physics, *Int. Journal of Theoretical Physics*, Vol. 38, 1.
42. Davies P., 1995, *About Time*, Simon and Schuster, New York.
43. Deutsch D., 1997, *The Fabric of Reality*, Penguin Books, New York.
44. Dubrov A.N., 1979, The interaction of biological objects with time and space, *Psychoenergetic Systems: The Interaction of Consciousness, Energy and Matter*, Krippner and Carlson (eds.), Gordon & Breach, New York, 115-121.
45. Gold T., 1967, *The Nature of Time*, Cornell University Press, Ithaca.
46. Goswami A.,1995, *The Self Aware Universe*, Putnam's Sons, New York.

47. Griffin D.R. (ed.), 1986, *Physics and the Ultimate Significance of Time*, State Univ of NY Press, Albany.
48. Halpern P., 1990, *Time Journeys*, McGraw-Hill, New York.
49. Margenau H., 1950, *The Nature of Physical Reality*, McGraw-Hill, New York.
50. Misner C.W., Thorne K.S., and Wheeler J.A., 1973, *Gravitation*, Freeman W.H., San Francisco.
51. Penrose R., 1985, Objective state-vector reduction, *Twistor Newsletter*, **19** 1-4.
52. Penrose R. and Isham C.J. (eds.), 1986, *Quantum Concepts in Space and Time*, Oxford Univ. Press.
53. Savitt S.F. (ed.), 1995, *Times Arrow Today*, Cambridge Univ Press.
54. Sklar L., 1985, *Philosophy and Spacetime Physics*, University of California Press, Berkeley.
55. Sklar L., 1995, Time in experience and in theoretical description of the world, *Time's Arrow Today*, Savitt (ed.), Cambridge Univ. Press, 217-229.
56. Toller M., *Variance of the quantum coordinates of an event*, preprint.
57. Wolf F.A., 1998, Timing in Conscious Experience, *J. of Scientific Exploration*, Vol. 12, 4 511-542.

Bergsonian Time and System Maturation

[1]JÖRG D. BECKER AND [2]IOANNIS ANTONIOU
[1]*Universität der Bundeswehr, München*
[2]*The International Solvay Institutes, Bruxelles*

1. TIME AND INFORMATION

Time plays a strange role in Newtonian physics, and also in ordinary quantum physics. Whereas an object's position and momentum, and functions of these two, such as energy or angular momentum, are considered to be *"observable"*, time enters as a *"parameter"*.

In general relativity things are quite different. Time does not exist a priory, but is *generated in processes*. Thus, a body moving along a path Γ from point A to point B in the 4-dimensional space (cf. Fig. 1), experiences a *proper time* τ given by

$$\tau^2 c^2 = \int_\Gamma g_{ik}(x)\, dx^i dx^k \tag{1}$$

$g_{ik}(x)$ the metric tensor which in turn depends on the motions of other bodies; c is the velocity of light in the vacuum. Only in very special cases time becomes something universal in the universe. If we assume that the universe is homogeneous and isotropic, i.e. it looks the same everywhere and in all directions - standard cosmological model -, then all observers falling freely with the cosmic "liquid" experience the same time which may then be called "cosmic time". There are three basic types of solutions to the Einstein equations: the Marxist-Leninist solution (matter is eternal), the Roman Catholic solution (the universe has a beginning and an end - point A and point Ω), and the existentialist solution (the universe is thrown into existence, it has a beginning but no end). One could also speculate about an

Studies on the Structure of Time: From Physics to Psycho(path)ology
Edited by Buccheri *et al.*, Kluwer Academic / Plenum Publishers, New York, 2000.

iterated Roman Catholic universe: the universe is born again and again - the Hindu solution (after R. Sexl; cf. Fig. 2).

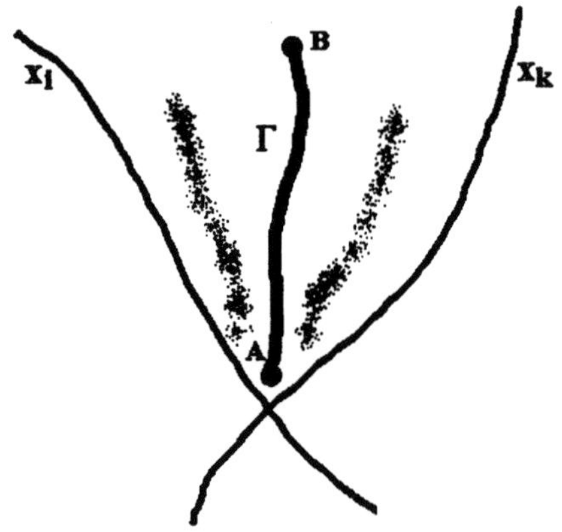

Figure 1. Body "moving" along a path Γ. Dotted lines: other bodies; $\{x_i\}$: coordinate system.

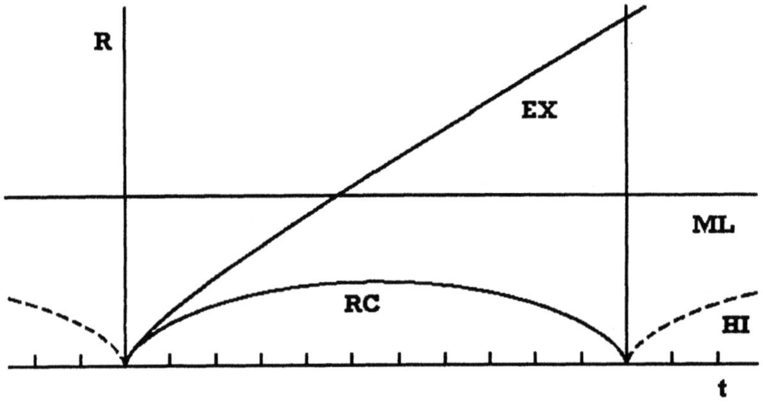

Figure 2. Possible types of solutions of standard cosmology: scale parameter R ("radius") of the universe as a function of cosmic time t. ML=Marxist-Leninist, RC=Roman Catholic, EẊ=Existentialist, HI=Hindu universe.

Under which conditions can we speak about a universal time? The point is that time is closely connected with other notions. Here, in particular, we have to take into account *information* - or, in a more restricted context, *entropy*. As long as information is *conserved*, as it is for a Newtonian planet, time looks like given *a priori*, as Kant would have put it, and we can relate it to the position $S(t)$ of the sun in the sky, or to the sequence of kings in a dynasty. If the sun moves to the right, we say that "time proceeds".

Time reversal is not a problem: if you travel to Rio de Janeiro you will notice that the sun moves to the left. However, as Hindemith[1] has argued, time reversal would not solve any problems, because for man it does not make any difference whether he is born first and then dies afterwards, or whether his life starts with death and ends with birth.

As soon as information *changes*, as it usually does in general relativity, time is no longer given a priori: time is a matter of maturation now. Eventually we have to define time to our convenience, or even to our pleasure, ultimately relating it to the position of a thought in a topological space of ideas. Bergson said, "For a conscious being, to exist is to change, to change is to mature, to mature is to go on creating oneself endlessly."

To be honest we must say that even in the case of Kantian time we need a record in a memory - i.e. an irreversible process - to compare the past with the present, to see that time proceeds[2]; as in the case of changing information something must be "standing", something must keep its identity. This points already to a kind of complementarity between Kantian time and change of information.

That we find it difficult to give up Kantian time may be due to the fact that time falls into the class of cultural items which are transmitted *formally*[3].

A more rigorous attempt to relate time to information has been formulated in Weizsäcker's ur theory[4,5,6].

2. TWO COMPLEMENTARY PICTURES OF TIME

As v. Müller[7] has explained, there are other notions related to time. With Kantian time we may associate classical logics. You may ask what time has to do with "timeless" logics; but time appears in Aristotle's second axiom, "It is impossible that the same applies and does not apply to the same *at the same time*". Likewise, as v. Müller[7] points out, strong causality (predictability) and the subject-object dichotomy are related to Kantian time; they form a *compound picture* which we call the linear-successive picture of time.

A second compound picture can be formed by Bergsonian time ("duration" in which life develops), dialectics (synthesis), structure formation (unpredictable, emergent), and ipsoflexivity ("subject" is part of the system and changes with it). Both pictures are consistent within themselves, but they must not be mixed: mixing leads to inconsistencies (Table 1).

We can easily relate the two pictures to the notions of *being* and *becoming* where being is related to constant information, $dS/dt=0$, and becoming (or decaying) is related to a change in information, $dS/dt \neq 0$.

Table 1. Two complementary pictures of time

Linear-Successive Picture of Time	Complex Picture of Time
Kantian time	Bergsonian time
Logics	Dialectics
Strong Casuality	Structure Formation
Subject-Object Dichotomy	Ipsoflexivity
Being	Becoming / Decaying
$dS/dt = 0$	$dS/dt \neq 0$

Furthermore, the two pictures are *complementary* in the sense of Bohr. As an expression of this complementarity we may see the uncertainty relation

$$(dS/dt)^2 \tau^2 = \mathcal{F}(\psi) \geq 1 \tag{2}$$

τ may be interpreted as the time horizon within which predictions make sense. This horizon is limited by the change of information in the system. Thus, there is a complementarity between *predictability* and *learning speed.* $\mathcal{F}(\psi)$ is a functional of the system state ψ, bounded from below. Even if this uncertainty relation still needs many investigations concerning its range of validity, its various interpretations, its precise form in a given system, and its consequences, it nevertheless provides an interesting formal link between the two pictures of time. (The uncertainty relation may also be useful for neural networks which learn permanently, even during operation: changing the rules during the game).

3. ENTROPY PRODUCTION FAR FROM EQUILIBRIUM

As an example for a change of information in a system we consider entropy production. It is closely related to the non-integrability of a system. The microscopic reason for it is that particles "remember" interactions; we may speak of "post-collisional correlations". (It is unknown whether other possible sources of the "arrow of time", such as cosmology or Murphy's Law, can be reduced to entropy production).

It is well known that entropy production cannot be formulated in the state space $\mathcal{H}$ (Poincaré-Misra theorem) but only on the space of density matrices $\rho \in \mathcal{H} \otimes \mathcal{H} = \mathcal{H}^2$ because entropy is a statistical entity. There are various ways of constructing a density matrix; for instance, if P_k is a family of orthonormal projectors on the states $\psi_k \in \mathcal{H}$ then

$$\rho = \sum_k \omega_k P_k \tag{3}$$

describes a statistical mixture in which the states ψ_k appear with probability ω_k. Because the sum of all probabilities must be 1 the trace of ρ must be 1.

To calculate post-collisional correlations directly is very cumbersome. Fortunately there is a very elegant way of calculating entropy production. Already Poincaré had noticed that non-integrability is associated with the appearance of resonances. A resonance may be viewed as a temporary "bound state" between two objects with a mean energy ω and a mean life time $\tau = 1/\gamma$ (Fig. 3). Contrary to the belief that we can formulate all physics with asymptotic (i. e. stable) states, here we need the resonances to compute physical effects.

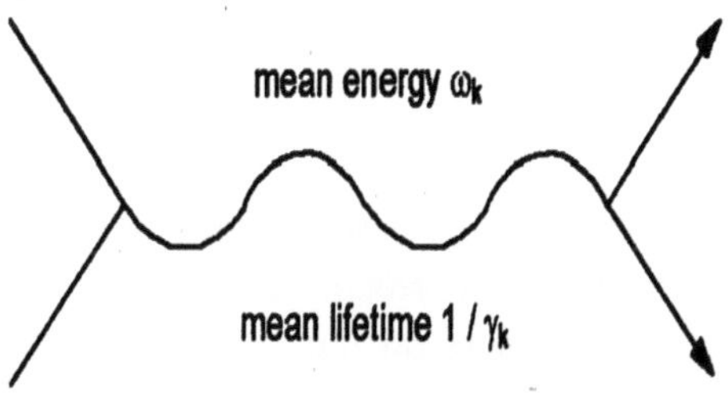

Figure 3. Resonance with mean energy ω and width ("uncertainty") γ; mean lifetime $\tau = 1/\gamma$.

The problem is that resonances do not belong to the state space $\mathcal{H}$. We have to enlarge the state space to a Gel'fand space triple (or "rigged Hilbert space"). Take a Hilbert space (e.g., the space of all square integrable functions) as a state space. Consider a subspace Φ of the state space $\mathcal{H}$, for instance, the space of all functions of the Hilbert space which are infinitely often differentiable. Consider the dual space Φ' of Φ, i. e. the space of all linear functional on Φ. We have

$$\Phi \subset \mathcal{H} \subset \Phi' \tag{4}$$

Now Φ' contains distributions (e. g. δ-functions), and resonances are states with distributive character. A weak topology is induced by the scalar product $< f, \phi >$, $f \in \Phi$, $\phi \in \Phi'$ necessary for studying the convergence of sequences.

To give a feeling for this situation, consider the Bernoulli map

$$x_{n+1} = 2x_n \bmod 2 \tag{5}$$

Time n is discrete. The corresponding time evolution of the density matrix (which is a function on the interval [0, 1]) is described by the Frobenius-Perron operator U_B:

$$\rho_{n+1}(x) = U_B \rho_n(x) = 1/2\left[\rho_n(x/2) + \rho_n((1+x)/2)\right] \tag{6}$$

The Bernoulli polynomials B_k on the interval [0, 1] are the eigenstates of U_B; they span the space Φ. The adjoint operator U^*_B possesses the eigenstates

$$f_k(x) = (-1)^k\left[\delta^{(k-1)}(x-1) - \delta^{(k-1)}(x)\right] \tag{7}$$

which are derivatives of δ functions and belong to the space Φ'.

We can extend this idea to Hamiltonian systems, constructing a space triple $\Phi \subset \mathcal{H} \subset \Phi'$, where Φ' contains the resonance states f_k with the corresponding complex eigenvalues $z_k = \omega_k + i\gamma_k$. The real part of z_k corresponds to the mean energy of the resonance, its imaginary part $\gamma_k = 1/\tau$ is the width of the resonance which is the inverse of its mean life time τ. Let $P(\omega)$ be the family of projectors corresponding to the continuous part of the spectrum; let P_K be the projectors on the resonance states f_k. We may then define an operator M by

$$M = \int \omega \, dP(\omega) + \sum_k z_k P_k \tag{8}$$

Finally, taking $M=\Lambda^*\Lambda$, entropy may be defined as

$$S(t)=\log \|\Lambda \rho(t)\| \tag{9}$$

One can show that we have $dS/dt > 0$ unless the system is in equilibrium. This describes the “arrow of time” in non-equilibrium physical systems.

Furthermore this approach gives us the basis for a theory of observation, memory, and measurement. A measurement is an irreversible process due to resonances between the measured system and the measuring system.

4. PRAGMATIC INFORMATION

According to Hartley, information is the *precision* to which a *given object* is located in a *given search space*. Shannon has refined this definition for the case that we have some a priory knowledge about the preferences of the system.

In general, both definitions are too restricted for many practical purposes. We want to give a definition which covers a wide area of reality, including arts.

From cognitive psychology[8] we borrow the idea that solutions to problems are to be constructed from a set of (given or found) building blocks and construction rules. Let us consider the set $\mathcal{H}$ of all constructs ψ which may be constructed from such a set. It may be, but need not be, a space in the mathematical sense; in any case, it will have some mathematical structure. Let us assume that $\mathcal{H}$ is equipped with a norm $\|\psi\|$ which measures the "size" of the construct ψ. This concept is quite general: constructs may be shapes, ballet dances, electronic circuits, symphonies, companies, thoughts, etc.

Let us assume furthermore that we have a *vision* V about the constructs to be constructed, and that we may formulate V as an operator on $\mathcal{H}$, $V{:}\mathcal{H}\rightarrow\mathcal{H}$, such that, if we have

$$V\psi^R=0 \tag{10}$$

we call ψ^R a solution, or a realization, of the vision V. There may be many, just one, or none such ψ^R. V has to be thought of as a decomposition or analysis operator: If the decomposition of a construct ψ^R is successful, nothing is left over. (A prototype for such an operator is an annihilation operator in quantum physics. Thus, the ground state Ω of the harmonic oscillator may be defined through the phonon annihilation operator a by $a\Omega=0$).

For a simple example consider the set of all shapes. Let P be the mirror operator with respect to a given point. Then $V=1-P$ is the vision operator which has any symmetric shape as a realization.

In most cases, for a given construct ψ, $V\psi\neq0$. We define *pragmatic information* associated with the vision V and the construct ψ as the *precision* to which the vision is realized by the construct ψ:

$$S=-\log\|V\psi\| \tag{11}$$

It should be stressed that the term "pragmatic" here refers to "estimates ... by its practical bearing upon human interests", not to "matter-of-fact treatment of things" (The Concise Oxford Dictionary).

We note that S is both subjective (because it depends on visions and ultimately on human values) and objective (once a vision has been specified,

S has a precise meaning). Furthermore it is distributed over various instances (it depends on the vision, on the construct, and on the environment).

Again, entropy turns out to be a special case of information, as it is already in the classical case[9] (cf. eqs. (9) and (11)).

We still need a change in S to speak about time. This change can be generated by learning rules. As an example we take evolutionary systems, or breeder controlled systems. An iterated cycle of mutations and selections generates a sequence of generations of constructs, and a pragmatic information associated with each individual.

There are two main differences between entropy production and evolution:

- $S(t)$ is not monotonic; even if on some time average $S(t)$ is monotonic there must be fluctuations such that evolution may escape local minima.
- The evolutionary dynamics cannot be formulated on the space of density matrices. Selection is a highly nonlinear process which can only be formulated on the space $\mathcal{H}^{\lambda}$ where λ is the size of the population.

Therefore evolution is quite different from "entropy production running backwards in time" (which is a nonsense in itself).

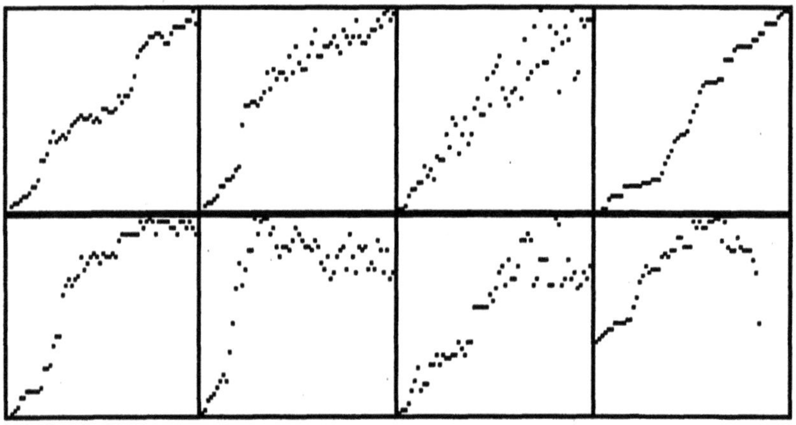

Figure 4. Bergsonian time as a function of Kantian time: simulations in evolutionary systems. Top row: various examples of successful systems; bottom row: stagnation and breakdown. Courtesy of Th. Schäff[15].

In the case of Kantian time we have used a pointer (the sun in the sky) to measure time. We can do something similar in the case of Bergsonian time. We can use maturity as a pointer: entropy in the case of a physical system, pragmatic information in the case of an evolutionary system. In the latter case, time becomes even subjective and non-monotonic. We may conjecture that feelings like "time does not pass"[10] are due to stagnation in Bergsonian time (Fig. 4).

5. ON SECURITY AND OPPORTUNITY

A world in which everything is predictable gives us security. Küppers[11] conjectures that security is the main concern of society. What can we expect from an unpredictable world? (cf. table 2)

Table 2. What if not security?

Linear-Successive Picture of Time	Complex Picture of Time
$dS/dt=0$	$dS/dt \neq 0$
predictability	unpredictability
security	?

In discussions on learning organizations, and in practical experiences, we have noticed that people talk about learning organizations ... as long as it does not concern themselves. At this point they cut off the discussion: ipsoflexivity is not permitted - we want to learn, but we don't want to change! Of course, we think that our identity is in danger ... but the question is whether our identity is related to our being or to our becoming.

If our identity is related to our memory then it is clear that this is due to irreversible processes, i. e. to becoming.

Bergson's starting point was his opposition to Laplace's demon: if everything is predictable there is no space for freedom and responsibility.

What then are complementary terms to security: insecurity, crisis? We rather have to associate security to stagnation, secure poverty, banality, with complementary terms like *chance, opportunity, innovation, richness, adventure, good luck*; or even *freedom, responsibility, democracy*, as Prigogine[12,13] has stressed out!

In practice we have to use both pictures: old ideas have to mature and to work, new ideas have to be created for future development. The question is rather how to find the political center between conservatism where new ideas don't have a chance, and permanent revolution where arbitrariness doesn't permit the formation of useful structures.

Senge[14] has made it clear that having mental models and common visions is crucial for learning organizations. Extending his point of view, we have identified some necessary ingredients for successful innovation:

- A multitude of values to which we can subscribe
- A common vision
- Adequate building blocks and construction rules
- An environment which encourages and rewards innovation
- A mental model of the complex compound of time

A most interesting aspect of our epoch is that it is possible to investigate metaphysical (and other philosophical) questions scientifically, in theory and in simulations. Ultimately, as we have learnt from Sartre, the values themselves are a product of a process of self-organization. Taking life as the mother of all values we may well come to the conclusion that the shareholder value alone will not suffice for the future, that economics as well as science are not life but have to be life's servants, and that consolation is more useful than competition.

ACKNOWLEDGMENTS

It is a pleasure to thank Prof. Rosolino Buccheri and Prof. Vito Di Gesù for their kind invitation to this conference, and to thank Prof. Ignaz Eisele for his generous support of this work.

COLLECTED SAYINGS

Time is what happens when the wheat matures. (Hopi wisdom)

L'entropia prolunga la vita. (Rossana Di Gesù)

Lots of bad TV prolong life. (Richard Walker)

Molti studiano come allungare la vita, quando invece bisognerebbe allargarla! (Luciano De Crescenzo[16])

Ne serait-il *(le temps)* pas alors le véhicule de création et de choix? (Henri Bergson)

Le temps est invention ou il n'est rien du tout. (Henri Bergson)

On peut dire qu'à l'équilibre la matière est aveugle, alors que loin de l'équilibre elle commence à voir. (Ilya Prigogine)

Jeder will etwas sein, keiner will etwas werden. (J.W. von Goethe)

Le monde se construit. (Pierre Teilhard de Chardin)

REFERENCES

1. Hindemith P., Hin und zurück, 1928, *Small opera*, Quoted after: Fünf kleine Opern der Zwanziger Jahre, Marstall-Theater, München.
2. Quadranti P., 1992, *Constructive realism*, Preprint, Geneva.
3. Hall E.T., 1976, *Beyond Culture*, Garden City (NY).
4. Görnitz Th., 1997, Naturwissenschaft als Informationstheorie, *Kreative Systeme*, Gerlach H., Becker J., Eisele I. (eds.), Harald Gerlach ICAS Press, München.
5. Weizsäcker C.F. von, 1986, *Aufbau der Physik*, Carl Hanser, München.
6. Weizsäcker C.F. von, 1995 *Einheit der Natur*.
7. Müller A.C. von, 1983, *Zeit und Logik*, Baur, München.
8. Osterloh M., 1983, *Handlungsspielräume und Informationsverarbeitung*, Huber.

9. Antoniou I., Becker J., *Pragmatic Information and the Misra Entropy Operator*. In preparation.
10. Pereira M.A., *The experienced time and psychical suffering: The perception of psychchiatric patients*, contribution to this conference.
11. Küppers G., 1999, Coping with Uncertainty: the Self-Organization of Social Systems, *Human and Machine Perception 2: Emergence, Attention, and Creativity*, Cantoni, Di Gesù, Setti, Tegolo (eds.), Kluwer Academic / Plenum Publishers, NY.
12. Prigogine I., 1980, *From Being to Becoming*, Freeman W. H., San Francisco.
13. Prigogine I., 1996, *La fin des certitudes*, Éditions Odile Jacob, Paris.
14. Senge P.M., 1994, The Fifth Discipline, *The Art & Practice of The Learning Organization*, Currency Doubleday, New York.
15. Schäff Th., Mustererzeugung und Selbstorganisation, 1999, *Studienarbeit UniBw*, München.
16. De Crescenzo L., 1984, Storia della filosofia greca: I presocratici, Mondadori, Milano.
17. Antoniou I., Tasaki S., 1991, Generalized Spectral Decomposition of Mixing Dynamical Systems, *Int. J. Quantum Chemistry*, **46** 425-474.
18. Becker J., 1994, Learning as gain of information, *The paradigm of self-organization II*, Dalenoort G.J. (ed.), Gordon & Breach.
19. Becker J., 1994, Entropy and Evolution, *Inside versus Outside*, Atmanspacher H., Dalenoort G.J. (eds.), Springer, Berlin.
20. Marsak L.M., 1961, *French Philosophers from Descartes to Sartre*, Meridian Books, Cleveland.

Time, Chaos & Entropy

ZACHARY JACOBSON
Ottawa, Ont., Canada

1. INTRODUCTION

Lewis Carroll wrote an apocryphal story-poem, *The Hunting of the Snark.* A hunting party searches for a Snark on an island where there is known to be one. The Snark is a magical beast—if you catch one it gives you everything you need and want.* There's a catch, to wit, the Boojum, also known to inhabit the same island. The Boojum is a magical beast as well-if you catch one you vanish forever. And, oh yes, the Boojum looks just like a Snark.

This is a metaphor for basic scientific assumptions. Assumptions can simplify, but incorrect assumptions can simplify incorrectly and allow you to prove untrue things. In fact, one false assumption can lead to proving almost any specific untruth†.

In the poem, it ended that the Snark they caught was a Boojum. Of course! Boojums are Snarks. As soon as you touch a Boojum, you vanish; that makes good the Snark's promise to care for all your needs. But, let me say, it is at least logically possible for a Snark not to be a Boojum--to be a beast that will care for you [or at least furnish a wonderful feast] and still not make you vanish, at least not immediately.

* This story seems to get better with the retelling. In the original poem, the Snark's only magical property is its wondrous flavour and texture when eaten; the other mythic properties have been written about and added subsequently. That does not affect this argument.

† The usual demonstration of this is that if you assume that 2 + 2 = 3 you can prove that you are the Pope (or anyone else you choose to be). If you don't know the proof already, try it. If it is too easy, use the same assumption to prove that all horses are grey.

Studies on the Structure of Time: From Physics to Psycho(path)ology
Edited by Buccheri *et al.*, Kluwer Academic / Plenum Publishers, New York, 2000.

A part of science is always to find those few assumptions that are unavoidable, are supported by extensive data, and which can explain as much of the world as possible. I wish to discuss the basic assumptions surrounding the Second Law of Thermodynamics and time's arrow. I will need to bring in the Heisenberg uncertainty principle and some aspects of chaos and fractals as well.

Why time is irreversible has been a puzzle in the bases of science at least during the 20th century. Quantum theory and Relativity, like Maxwell's and Newton's laws before them, are independent of time's direction. That is, the laws describing the physical interactions among particles and planets all seem to work equally well in both time and reverse-time. Our everyday world is made up of molecules that are made up in turn of particles whose individual behaviours are explained completely by laws that do not care which way time goes, yet time's direction is clearly very important in that same everyday world.

Einstein treated time as a dimension *comme les autres.* That works very well for Relativity and Quantum theory, that is, for virtually all of modern basic physics and chemistry. But is time a dimension exactly like the three spatial dimensions? You can move readily in either direction of any of the three dimensions in space but you cannot go backward in time. You cannot stand still in time nor can you move forward at any other than the "standard rate", relativistic effects aside. It's just too bad if you don't want your birthdays to approach as quickly as everyone else's or if you want to meet Shakespeare face-to-face. Moreover—so far as we can see—we're all going the same way at once, in lockstep, and there is only one instant extant at once [at least locally].

I suggest that chaos, information and entropy come together to yield irreversible time. Rather than look at the small quantum world building up to the everyday world, this is the result of the large world impacting on the small one. Trajectories in the larger world, getting smaller and smaller in chaotic mixing--as cigarette smoke disperses into still air--convert to entropy when they are too small to affect the cloud of quantum uncertainty associated with the molecules they impact upon. If you take this view, you can account for time's arrow and the development of entropy--two basic birds with one stone.

Chaos and complexity have been developed as mathematics over the past few decades, most rapidly in last ten years[1]. Concepts now established include the nature of fractals-lawful structures that are infinitely detailed, appearing identical at any magnification. The now-familiar pictures from the region just outside the Mandelbröt set are fractals. Fractals appear in Nature, e.g., weather patterns, coastlines, tree skeletons, river courses, and many natural landscapes.

Look at a Mandelbröt picture on a computer screen. The smallest jitter of your point of view has only the tiniest effect on the picture you see-possibly invisible. But at increased magnifications, the effect is profound-the structures have all changed totally. You might as well have moved to another planet. This effect, particularly when future predictions of chaotic but lawful events are involved, has become to be called the butterfly-wing effect. The reason for that name is a true story that sounds apocryphal. Konrad Lorentz mislaid some iterative, complex weather calculations in an epoch before computers were as powerful and plentiful as they are today. Lorentz replaced the computations using a hand-operated calculator, rounding off to two decimal places for convenience. He expected that there would be no important change due to the rounding error. Forecast weather remained the same for a time [during several iterations] but soon it diverged dramatically from the earlier predictions, which had been rounded off after a higher precision. The joking way it was described was nevertheless quite accurate-motion of a butterfly wing in Hong Kong [say] today may affect weather patterns over New York within days or weeks.

In their thoughtful and authoritative book, Coveney and Highfield[2] suggest that chaos is intimately involved in the arrow of time, although they are vague about the mechanism. It is obviously tempting to attribute time's arrow to butterfly-wing non-computability-at-the-limit as that mechanism, but I doubt it. More later on.

2. QUANTUM UNCERTAINTY

The Heisenberg uncertainty principle began life innocently enough. If you want to know where a particle is, you could shine a light on it. The photon that impacts it will indicate exactly where the particle was when it was hit by the photon, but the photon will impart a momentum to the particle in so doing and that will be unknown. The unknowability is such that the multiple of what you don't know about position and momentum (itself the product of mass and velocity) will at least be equal to Planck's constant.

$$\Delta p \, \Delta v \geq h$$

Note that, as things get larger--i.e. as they have more mass--the uncertainty in momentum becomes much smaller. At the level of an elephant, the uncertainty of position and velocity are vanishingly small, but they are considerable at the mass of an electron. Mass and velocity are termed a complementary pair of properties, and there are many other such pairings. A classicist in physics [if you could still find one] would adhere to something like the causal, photon-knocking, explanation of each property pairing. Einstein and de Brőglie favored that view. So did Schrődinger-that's why he proposed the famous cat-in-the-box gedanken experiment as a disproof-a variety of *reductio ad absurdum*-showing how the Bohr's interpretation of quantum theory led to an impossibility. But that backfired in one of the most celebrated cases of a scientist eating humble pie in the 20th Century.

3. THE SOFT AND HARD INTERPRETATION - COPENHAGEN SCHOOL

Bohr, Heisenberg, and others in what came to be called the Copenhagen school, thought differently than the classical description. If it can't be measured, Bridgman had said, a quantity does not exist! Nonsense said Einstein-if an electron is moving toward you from infinity (that could be as far away as a meter) you may choose which property you will measure to an infinite precision. So it has both properties, no? Well, no, as it turned out.

Electrons, photons, and all the other entities have wave properties as well as particulate ones. Now if you pass a wave through two slits, it produces a standing wave pattern on the other side-as the emerging forces reinforce or interfere at different positions. If a single electron is arrayed to pass through only one slit left open in the apparatus, it behaves as if it went through the one slit. And if another slit is opened, surely it should not matter- the pattern would be the same as if the electron went through one or the other slit-but not both, right? Wrong! Opening the two slits simultaneously makes even single electrons behave as if they have passed through both slits-the interference fringes show up undeterred by the human imagination that the electron is a little ball in only one place at a time. The same thing holds for photons.

4. THE DIFFERENCE BETWEEN QUANTUM UNCERTAINTY AND MECHANICAL UNCERTAINTY

Quantum uncertainty principle makes me very uncomfortable. It seems to negate the notion of local reality or causation. Bohr said anyone not shocked by Quantum theory did not understand it; this is what he had in mind.

Butterfly-wing unpredictability or chaotic unknowability is quite different than the unknowability set out by the Heisenberg uncertainty principle. In the chaotic realm, unpredictability is identically the commonsense, classical interpretation of unknowability that Einstein favored. That is, all action is lawful and described by equations that are knowable, at least in principle, and computable if only you had sufficiently precise data and enough computing power, neither of which break the laws of physics. But under the Copenhagen interpretation, unknowability is elevated to the stature of a law because the quantities *simply do not exist* until they are measured. The two-slit experiment and the Bell inequalities are physical evidence that the Copenhagen school is maddeningly correct at small dimensions, and quantum randomness is now accepted generally as a given law of nature.

Let me be precise about the peculiar nature of quantum uncertainty. Tossing a coin--the standard example of a random event given in most probability textbooks--is a causal and lawful event, but not truly a random one. If you know the precise starting conditions, all the forces, momenta and positions as you toss a coin, you could predict heads or tails. Heads or tails is not predictable only insofar as you do not know the starting conditions with sufficient precision [perhaps you intentionally refuse to make the needed measurements!]. This does not affect the study of practical probability and statistics for the business world or the social science laboratory; but I think it has a profound effect on other thinking; more about that later on, too.

One basic thing that chaos does is at least to start to account for the emergence of complexity and structure out of simpler substrates. The most basic descriptors of the Universe are Relativity and Quantum theory, which replaced [encompassed really] Newton's Laws and Maxwell's Equations as the legacy from previous physics. From those beginnings, it seemed possible to construct first our knowledge of matter and the heavens (physics and astronomy), then chemistry, then biology, then physiology. But each of these steps represents an orders-of-magnitude increase in complexity; that seems anti-entropic, but this added complexity is due to injections of proportionately large doses of energy, driving the systems far from equilibrium; they are not allowed to fall into a steady state. Even so, how complex structures emerge from simpler beginnings is always puzzling. The worst problems involve genes coded on DNA: it is understood how DNA codes for specific proteins, but how do they end up in structure. Why does a dog not look like a jellyfish, say, or a lizard, made up of canine proteins sent down from the DNA? Chaos and complexity theory give the hope that answers may be available.*

Gleick and Gutzweiller among others separately note that there has been no success in developing a quantum theory of chaos as yet. That is, and

* Given the information expected from the genome-mapping projects ongoing, of course.

current attempts notwithstanding, it has proven difficult to provide a generally accepted bridge between lawful events at the quantum level and chaotic events at the same level. This is certainly going to be a profound fact if it remains impossible to provide that bridge, but it seems to me too early in the development of chaos and complexity as disciplines to say if it will be true.

The important thing about chaos, Coveney and Highfield suggest, is that it inserts the directionality of time into the mix of fundamental properties. However, like thermodynamics, chaos still applies only with regard to larger-than-Quantum systems.

There is, in other words, an important property of the world that is not subsumed by the basic scientific theories accepted during the bulk of the 20th Century. The solution must seem painfully clear to anybody who isn't a physicist; the theories aren't complete; time's direction *is* fundamental. However, and maddeningly, the theory seems to be complete without it. The math seems to need (nay, permit!) nothing else, but that the effects and interactions and phenomena all seem to work equally well in both time and reverse-time. Yet our world, obviously time-dependent, is made of the totality of those particles whose individual behaviour is so completely explained at a microscopic level by time-invariant physics.

There are some basic questions that are not included in the ladder from quantum physics up to the workaday world and the direction of time is high on that list. Quantum theory and Relativity, like Maxwellian and Newtonian laws before, are independent of the direction of time. That is, as noted earlier, the laws "don't care" if time runs forward or backward.

Chaos is an exception to the expectation that particle physics accounts for the rest of things-entropy is another. Thermodynamics always speaks of accretions rather than individual particles, so it still leaves a gap in the chain of theory that is generally accepted as more troubling than other, more mundane scientific lacunae. There are many ways to state the Second Law of Thermodynamics, here is a simple one: left to its own devices, heat only moves from a hotter body to a relatively cooler one.* Difference between the temperatures of two bodies has the ability to do work, and it is the major way that stored energy is made useful in human and natural economies.

The Second Law cannot be derived from processes involving individual particles because heat and temperature are properties of collections of particles. All particle processes are reversible-that is they would work

* Actually, that definition seems circular. It is true nevertheless; one needs to define the heat in a body without reference to the body's temperature. The usual examples that do not involve heat imagery are, it is easier to get toothpaste out of the tube than to get it back in, and it is easier to stir milk into coffee than to get it back out.

identically if time were reversed. "Reversed relative to what", you could ask. And the only answer you were likely to get until 1990 would have been, "relative to heat transfer". That is, that was almost the only answer available before the development of the theory surrounding chaos seemed to give us another.

Some[3*] would have said "relative to the Big Bang", but they would have had serious trouble relating that long-past event to the present question of reversibility-irreversibility. I find it unconvincing because, while it gives an anchor to the times of events, it does not seem to have local effects; it is absent from the two "basic" theories to boot, like chaos and the Second Law.

Weak nuclear interactions are known to be time-irreversible. The first direct observation was reported at the Palermo meeting[4]. However and notwithstanding the above, it is so weak and remote from other particle interactions that no-one thinks that it can possible be the source of time's arrow. [Coveney and Highfield and the Palermo authors take pains to point this out.]

If the cosmological argument and the weak nuclear explanations were not completely convincing, we would have been left with the heat bleeding irreversibly from warmer to cooler bodies as the only other plausible local source for time's directionality. Once the bodies have reached the same temperature (equilibrium), no more heat transfer goes on and that means that the original temperature difference is gone now and can do no more work. The process can be reversed from outside, of course, through things like refrigeration technology, but not without expending more heat differential (hence more usable energy) than it makes available.

Even though the Second Law cannot be derived from one-on-one particle interactions, it is clearly true, and rational scientists accept it as a fact of nature, at least a fact of the nature of systems or entities made up of enough particles to require statistical treatment. As heat moves, entropy is created.

* The best-known current proponent of this notion is Stephen Hawking. He has a lot of company. Most cosmogonists and astronomers accept this interpretation. Surely the Big Bang gives an anchor-nothing happened "inside" the Universe before that. There are some fascinating speculations possible, separate from the present argument, which are: if time relates to the expansion of the Universe, it ought to reverse should the Universe start to collapse. Of course, if that happened, we might not notice it, because we would still see things happen in the same order relative to *us*. Hawking himself is convinced that this is not the case, but I have not heard the arguments. For example, if time did reverse in a shrinking universe, would it appear to reverse, or would we still feel that we were living and aging in the direction of larger universe? Would the direction of turbulence reverse? If that reversed, i.e., went from smaller to larger in positive time, then apparent time--to humans--would reverse so completely that we might still see the Universe appearing to expand! (How this would affect the cosmological red shift. Would it appear as a blue shift?)

Entropy is an ineffable quantification of disorder, a quantity that always increases in closed processes involving statistically large numbers of particles. Recall that temperature itself is usually only meant as the average speed of internal motion of a large number of particles in a solid body or a body of liquid or gas. Recall also that there is no closed-form solution for the three- or higher-body problem. The familiar chemistry textbook picture of covalent bonding uses approximations drawn from the two-body solution, which seem adequate to the task.

Boltzmann developed a description and theory of entropy that applies in a perfect billiard-ball universe*. But that is an unsatisfying account for time's arrow because it is made of individual interactions that are reversible in principle but that reversal would reverse the entropy arrow with it. As well and as before, the measure of entropy is only meaningful in a statistically large number of particles.

Information equates to the negative of entropy in most common scientific interpretation†, and that is interesting in computation and economic theory. Thus, there is one more limit at the foggy edge of knowledge: if the universe is closed, i.e., if there is no supply of negentropy [order] in the universe, the whole cosmos will run down eventually. And no matter how good brains or computers get, there will always be a minimum energy cost to compute a bit of information using classical methods‡. That makes two new limits in the "knowledge fog" that we have noticed since mid-century: Advancing entropy asserts that we could never finish computing the course of events in the Universe before they occurred, and the butterfly wing assures us that we would be quite wrong in chaotic realms near the start of the computations and long before thermodynamics wrote *finis* to our work.

No one currently suggests black holes as sinks that would capture entropy, thus saving the Universe from a heat death. I suppose it is just barely conceivable-but there would need to be a mechanism to enable them to capture proportionately more waste energy than everything else they

* Boltzmann's treatment underlies Taylor's demonstrations, discussed later.

† Brouillon calls the negative of entropy "negentropy". It's a useful term because negentropy means order, or information, or heat differentiation, whereas entropy means disorder [as an absence of order or a lack of useful information] and heat diffusion.

‡ Quantum supercomputing now makes the promise that computation can be done below that limit by taking results out of parallel quantum states. The effect is about the same as noting that a beam of light moving through a snowball illuminates all of the possible paths through it-if a beam of light, then in the quantum world, also a single photon. Landauer, who described the computational floor in the first place, disagrees. He objects that the uncertainties in such a computer would sum to an unstable situation and the results would not be more trustworthy than an equivalent classical computer; he calls it the no-free-lunch principle. Experiments are under way with linked systems of several qubits; the only report I have seen a this writing tends to support Landauer.

capture. In fact, Stephen Hawking has suggested just the opposite-black holes capture and destroy negentropy (information)! Hawking suggests[5] that black holes may return their captured matter to the universe ("evaporate"), but only after losing all their historic momentum and positional information. I would like to build on that speculation a little farther on in this paper-and with a twist.

Individual interactions among individual particles do not sum to an explanation of the Second Law; this troubles me and makes me uncomfortable in drawing the conclusion that advancing entropy by itself is where the fundamental arrow of time develops, and I have company in this. For one thing, it is possible to run the standard equations backward in time in such a way that, given the position and momentum of every object in the universe (with perfect precision), you can extrapolate backward in time to "postdict" everything, and find out the causes of things-the usual example is, did Lee Harvey Oswald truly kill President Kennedy and did he act alone?

Chaos theory appears to Coveney and Highfield to offer a hope for a mechanism underlying time's arrow, however. The butterfly-wing phenomenon says that even when the world can be predicted in its future state with perfectly true relations (usually differential equations applied recursively over small temporal increments), accurate prediction is always impossible. It requires infinitely accurate knowledge of the present state-impossible even under the soft interpretation of the quantum theory-as well as virtually infinite computing resources. *Voila*, Time's Arrow, perhaps.

And perhaps not. It seems unlikely to me that soft uncertainty (the mere absence of knowledge) is enough to get time's arrow from the Butterfly wing. What remains clearly true is that we know the past and we cannot predict the future.

I believe that even the most elegant attempt to get time from chaotic [butterfly-wing] unpredictability alone is fatally flawed. Consider the Butterfly Wing observation. If you round off the quantities to two decimal places, the weather forecasts diverge after a time from those made with a greater number of significant digits. Well, the differential equations at the base of the forecast model work backwards as well. Lorentz could have as readily run his calculations backwards as forwards, and the weather "postcasts" and would have diverged* just as surely, from the same rounding- off, but users seldom request a weather postcast. Deriving past weather from present conditions would be useful as a calibration tool for forecasting differential equations and computational techniques, but records are normally kept of what the weather actually was hours and days and years ago. We see that the butterfly wing is as bi-directional temporally as the

* It would be trivial to extend this thinking to small finite numbers of objects in chaotic states.

orbit of a planet! A Feynmann diagram of the uncertainty due to chaotic uncertainty in space-time even looks like a butterfly's wings (see figure).

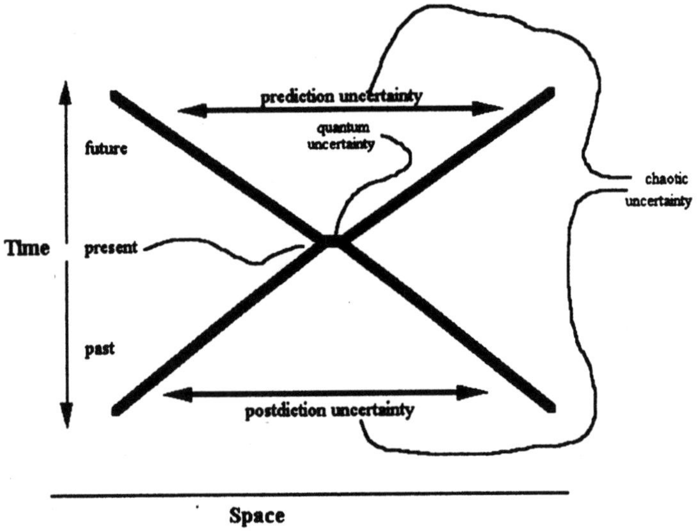

Figure 1. Uncertainty due to chaotic ignorance.

Records and record-keeping apparatus are totally outside of Butterfly wing unpredictability and "un-postdictability". Yet records are what make prediction different from postdiction, not chaotic relations in recursive differential equations. The capacity to record something seems totally removed from the sequence of math and computation that makes prediction and postdiction possible. But records only work one way in time-from past to future. So it is hard to imagine how you could write records if that process were not somehow outside the province of time-reversible physics. Thus it seems surely that the naked butterfly wing cannot be the origin of time's arrow. Recording uses and stores information, usually in ways that cannot be destroyed by readout in anything like the way that measurements destroy quantum scale information.

If the Second Law of Thermodynamics, with its monotonic increasing entropy in closed systems, truly lies outside of particle physics and relativity, then that adds entropy to the number of assumptions needed to account for the universe*. If chaos also remains beyond the ken of quantum physics, then it seems to represent a further complication that must be accounted for. Relations of chaos and thermodynamics are intriguing but puzzling. You

* Even after a realistic GUT can be described and confirmed, unless it also includes an explanation of the Second Law in terms of one-one interactions of single particles.

cannot make directional entropy fall out of something bi-directional, like the butterfly effect; further, reversing that and trying to bring the butterfly wing out of entropic reactions would involve throwing away a perfectly good time-dependence. How profligate that would seem.

All is not lost, however. The butterfly wing is not the only thing in the chaotic realm that suggests time-or more precisely time's direction. Many advancing processes get progressively and unstoppably more messy and chaotic with advancing time. Those processes are reversible in principle, though not in computational fact; I suggest that at the limit they become irreversible *in principle* and that is the exact point at which information dissipates into entropy and time's arrow arises.

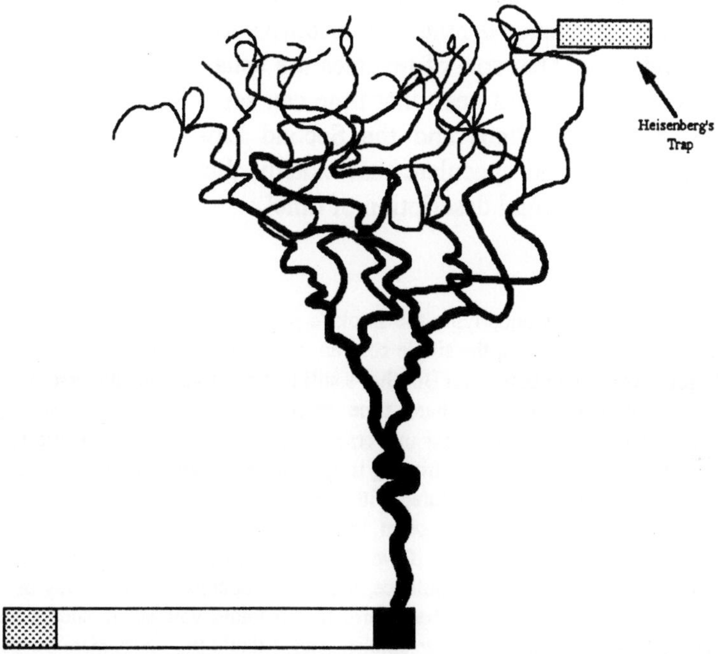

Figure 2. Cigarette smoke: what happens to the heat vectors when they fall into Heisenberg's trap?

Look at a lit cigarette [if you still can find one] in an ashtray in a still room. The smoke rises smoothly in a single-line laminar flow for a short space, then the path wobbles and becomes unpredictable and convoluted in a short space, even though the flow is still laminar and lawful on a microscopic scale. Now, the column keeps wobbling and moving at a smaller and smaller scale until-at the limit-the smoke is diffused molecule-for-molecule among the air molecules. Then the smoke is dispersed in the air; retrieving it would be more difficult than allowing it to disperse was.

If the smoke is hotter than the air (and it obviously starts out hotter), its heat differential relative to the air becomes dispersed in the air molecules' Brownian movement as well. At least here, I suggest, chaos has reduced itself to the molecular level and it has acted as a mechanism underlying the Second Law[*].

My modest proposal is that the sources of things run as follows: some chaotic events occur progressively in time, and they are at least part of the mechanism for the Second Law. In other words, chaotic events too small to be resolved due to Quantum uncertainty are the source for the growth of entropy, and time arises at just that point. Time is not relatively primitive to turbulence, but turbulence is relatively primitive to at least some realizations of the Second Law. That sort of reverses the usual way of looking at these things, where the hope has been that time can seem to be a result of one or the other or both of chaos and the Second Law, acting independently. Coveney and Highfield speculate that exactly this reversal of thinking is required for a fundamental description of time[†].

[*] The differential equations underlying the smoke's progressively convoluted rise allow you to project backward a toward the single column, to be sure, before the end of the process. (That's going from time tx to $t_{x-\Delta x}$) But that's still in a positive time universe. If you found yourself in another universe where time really did run backward, so would the differentials. Going backward in our universe's terms in the equations would run from tx to $t_{x-(-\Delta x)}$ which is to say, forward from t_x to $t_{x+\Delta x}$.In that reverse-time universe, then, the smoke might also get more convoluted with increasing time (as measured by an observer in that universe), even though it seem look to us as though the column developed downward from a cloud of smoke if we could see across from our universe. That's a convoluted way of saying that turbulence, too, might operate the same way backward in time, if you could make time run backward. If you found yourself in such a world, you could not tell it by watching cigarette smoke. And if that's the mechanism for the Second Law, you couldn't tell it by tracking heat or information, either.

[†] They write, toward the end of their book, *"At the present time, most scientists advocate the reductionist view. But ... we need to consider the universe in its totality and so adopt a more holistic outlook. ... A new theory is needed, one that will provide a deeper understanding of time. ... [C]learly what the new theory should do [is to:]*

- *banish the singularities that undermine Einstein's relativity;*
- *end the argument about what the act of measurement does in quantum theory;*
- *definitively outlaw time travel, auto infanticide and white holes; while*
- *rendering nugatory arguments based on subjectivity and illusion."*

My suggestion here meets at least three of these requirements—I am not able to assess the singularities. A few pages earlier, the writers allude to ongoing work by George, Henin and Prirogrine [GHP], which started *circa* 1970, dividing motion into kinetic (i.e., reversible) and non-kinetic sub-dynamics. My thinking is not totally new, clearly. What I do here is to suggest a mechanism that accounts for their piece of "lateral thinking"; viz.,

The suggestion that chaotic processes are at least part of the mechanism underlying the Second Law may be able to replace one different assumption, and perhaps two. If the one new assumption only replaces one older one, it may not be very interesting. But if it should replace two, then that *is* interesting. That depends on whether or not you can fish the Second Law out of developing chaos theory without further complicating it. I find it annoyingly elegant that the two puzzles, time and the second law, might be joined exactly by the third puzzle-Quantum uncertainty—the one that I find personally most nettlesome.

Coveney and Highfield posit that the law of increasing entropy is the local root of time's arrow, and they suggest that the Second Law be accepted as fundamental in the same way that Quantum theory and relativity are, and that, essentially, is that[2]. They are a little more ambiguous about the mechanism, but they clearly feel that chaotic mixing is at the root of that connection. They cite a proof by Prirogene and Courbage to the effect that "... if an entropy-like quantity exists that increases with time, then the reversible trajectories [of molecules in a gas] cannot be used." I believe they are quite right, and I suggest a mechanism.

5. A CLOSER LOOK AT TIME & TURBULENCE

Consider the plume of cigarette smoke again. The smoke rises smoothly until it wavers, then it wavers and tumbles faster and faster. The tumbling is a fractal and has no theoretical limit; you couldn't tell what magnification-scale you were looking at just from the grain of the picture. But when the scale gets smaller than-say-half the average diameter of an average molecule of the surrounding air, the molecules of smoke are mixed as thoroughly as they will ever be. Still going ahead in (forward) time the differential equations governing the tumbling forces are continuous and so the vector gets smaller and smaller; in few more orders of magnitude, it approaches the size of the Heisenberg uncertainty on the locations of electrons in the atoms of the air. I don't think that will have much effect yet, but in several more orders of magnitude of shrinkage, the turbulence vectors (carrying the energy differential due to the heat that started out in the smoke) approach the size of the quantum uncertainty of the locations and momenta of the air molecules.

And now, it matters profoundly whether you take the hard or soft interpretation to quantum uncertainty. Under the soft (einsteinian)

---

trajectories get lost in the transition to the Quantum realm. They do not get lost, exactly-the information which they represent reappears as diffuse heat-i.e., entropy.

interpretation, this is merely the place where you could measure no further tumbling vectors without impacting the molecules enough to change the result of reversing the calculations back up the scale of size. That would not be especially meaningful, because the postdictability would still be there, if only you knew the trajectories and masses and other properties to sufficient detail.

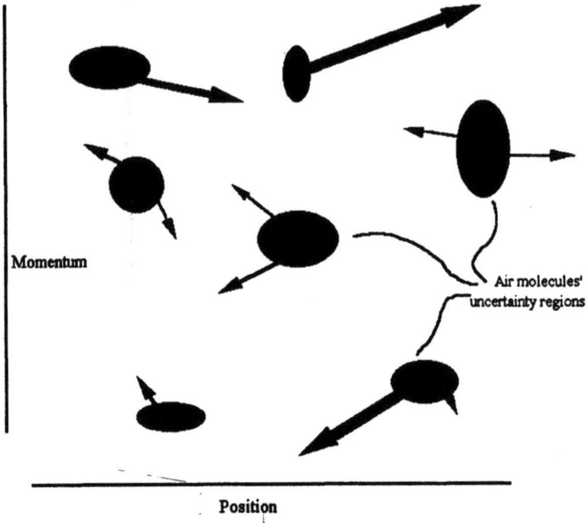

Figure 3. Heisenberg's Trap: Big vectors easily push air molecules around. What do the little vectors do?

But under the hard interpretation, favored by Heisenberg, Bohr, and Bell (holding his nose), as is essential to e.g., tunneling electron microscopy and quantum cryptography*, and as is favored by almost every contemporary physicist, there has to be a fundamental change at that scale. The vectors' information must be lost at that scale, irretrievably and time-irreversibly. The energy that carried the information is not lost of course; it gets diffused among the motions of the air molecules as waste heat. But the pattern of that energy is gone† and any useful information the pattern carried is gone with it. Even if time were reversed, you could not reconstruct exactly the same

* The hard interpretation seems to me to be forced by the double-slit data and the Bell inequality. Oh well, it may feel awkward but at least we can use it for something. Hence the technologies that use it, and hence this interpretation of the Second Law, depend on it.

† The energy is diffused among the motions of air molecules, but the processes are statistical and random at just this point. Exactly what motion in what molecule comes from the plume of smoke cannot be predicted (or postdicted, rather) any longer, according to this formulation, once the Heisenberg limit is reached.

column of cigarette smoke that you started with, though you presumably postdict a slightly different column.

At least for turbulent, convective heat transfer, a visible source for time's arrow seems to emerge as the chaotic, lawful, motions of heat knock against the uncertainty law. This does not quite say that turbulence and chaos cause the Second Law and are the only source for time‡; it also needs the standard interpretation of Quantum theory. Heat transfer by radiation does not go through a liquid or gaseous chaos, but the heat is already in quanta (photons), and thus it already has gone through the Heisenberg demon that, even if it does not remove energy, it obscures or removes it's local detail.

Which one-removes or obscures-depends if you are a classicist or if you follow one of the normal modern interpretations of quantum theory. As we see, it is not a trivial distinction. The rigorous interpretation of the uncertainty principle-that quantities paired under uncertainty do not exist until measured-implies that this is a source of time's arrow if the information acts as I described. Einstein's interpretation of uncertainty-that the quantities exist even if they cannot be measured-reaches no such conclusion. The fractals just keep getting smaller and smaller indefinitely, and are indefinitely reversible. "It's turtles all the way down." I'm sure Einstein would find it unsatisfying if things worked out the way posited here, but at least he'd finally know *why* God plays dice with the Universe—it's to give it time.

Stripped to its simplest, the suggestion admits of two metaphors:

1. There is "friction" at the boundary of information and quantum uncertainty; or
2. The smallest impacts of molecules carrying the dispersing heat upon molecules in the cooling medium "collapse" the wave function of the cooler medium's molecules-what the Copenhagen school called a measurement.

   Either way*, the information carried by the heat differential simply gets lost†.

‡ The present suggestion regarding time's arrow is interpretation-dependant, to be sure. Under Einstein's classical interpretation, for example, uncertainty relates to measurement but not to local reality; hence there is no true irreversibility and no obvious source for time's arrow. Similarly, some quantum physicists adopt the position that the Universe is described perfectly by a huge Hilbert matrix, which is perforce invertible. The adherents to the "Church of the Larger Hilbert Space", as some of the adherents jokingly call themselves, likewise find no source for time's arrow here. I know of know empirical support for this belief, but farbeit from me to criticize another's religion.

* I find the second to be the more appealing. Brassard suggested it in a conversation during 1996. It may have also appeared in the report of a Russian conference on temporal physics held about 1994, noted in the ISST Bulletin about 1995. I cannot find the source again, regrettably.

Note that this is a purely *local* source. It does not refer to the start of the universe, or to its expansion; if those things impact local turbulence, this may be a door through which they affect or effect time's arrow. I do not know of any other likely mechanism by which the big Bang and the expanding Universe could do that, however•.

There are two antecedents for this notion. Landauer[6,7] proved that all computing is reversible-and does not need the expenditure of energy-except those operations that require *erasing or losing information.* A few years after that, Charles Bennett[8†] used that result to show that any Maxwell's demon must use at least as much energy as it gains when sorting out hot (fast) gas molecules from cold (slow) ones, owing to the minimum information manipulation it must do in the process‡.

As seems appropriate, the foregoing are postdictions from the Heisenberg-trap explanation for entropy, and that for time's arrow. That is to say, they would be reasonable predictions if the sequence had been different. I can offer a speculative prediction. This model shows progressive chaotic mixing to be the source for entropy, thence time's arrow. If you could create a closed artificial system where mixing ran the other way in time, i.e., from small to large columns of cigarette smoke, or cream out of coffee, then time would run backward in such a system, as well°. If you placed yourself inside

---

† Time costs trajectory. To the extent that trajectory carries information, then, time costs information. To the extent that information has economic value, you could make this argument say that time costs money. Note that this is *not* an equation-viz., time *is* money. According to this picture, all the money in the world cannot truly buy back even one minute.

• The heat from the lost trajectories diffuses among the microwave background blackbody radiation.

† These last references, to Landauer's and Bennett, are antecedent to the turbulence-entropy-time suggestion because they were known beforehand, but not by me. With hindsight, they would have been reasonable confirmatory tests to suggest if they had not been previously established. I am indebted to Gilles Brassard for telling me about Bennett's and Landauer's work.

‡ Coveney and Highfield, citing Prirogene before them, disagree with the standard interpretation that information is identically the negative of entropy. It does not matter to the argument so far: the same notion would apply if the vectors were modeled as only the impact of molecular motions due to the heat of the smoke, independent of any information that might or might not carry. However it might be significant if one wished to develop this model further, into macroeconomics, say. Landauer is in no doubt about it. The title of his 1991 article in *Physics Today* is "Information is Physical." His text left no doubt that he specifically means that a computational loss of information is identically an increase in the computer's entropy. Unfortunately, he did not offer a derivation of that point.

° This assumes you could make the observation without disturbing the system and destroying its "closedness". I do not believe you could build such a system; it would be logically about the same as building a perpetual motion machine.

of the system, you would see mixing move in the normal direction, of course.

A question still remains for this notion, concerning conductive heat transfer. Imagine a cool aluminum cube placed in face-to-face contact with a hot one. The heat transfer to equilibrium is well modeled as smooth and unchaotic, yet the Second Law holds. One would have to recast the interface model as chaotic and tempestuous at the atomic level, where the hot aluminum atoms from one side knock on cooler ones on the other side and much of their imparted motion is lost in the uncertainty trap before it can be reflected back into the hotter cube. I find that argument a bit tortured, but I cannot see anything wrong with it. It merely seems to remind me that chaotic and fractal mixing turbulence are central to this notion mainly as an existence proof that there is a question to be addressed at the fringe of the quantum domain; anything that reduces the information in a hot body to the level of the individual particles that it must impact upon in the next instant runs into the same situation.

The one critical assumption is the Copenhagen interpretation of the position and momentum of the particles that carry the motions of the heat. If those particles truly do not possess precise momenta and locations until some energy impacts them sufficient to collapse their wave function, anything much less than that will have no effect and will disappear as a differentiated quantity. It will reappear somewhere, as it sums statistically with more of its fellows-in-limbo to impart one quantum of motion to one particle somewhere, so that the energy is conserved as heat, but undifferentiated as to its sources[9*].

Bennett considered this model at my request; he objected that it has a flaw because there are no trajectories calculated in quantum domains. I am not sure that this is a solid objection, as seen from a perspective outside of his discipline.

The standard picture includes these basic assumptions:

- Uncertainty as a basic principle, as are
- Irreversible time,
  - Chaotic mixing (arguably),
- The general increase of entropy in closed systems, and
- Trajectory in the macroscopic world

(while objecting that we may not need it in the quantum world).

The current suggestion has the elegance of fewer assumptions. It now assumes as basic:

* A small stimulus will coalesce a wave equation some of the time, with a statistical probability sufficient to assure that the kinetic energy shows up as heat. But the information still disappears. Entropy has no homologue in the quantum domain.

- Uncertainty,
  - Chaotic mixing (as well), and
- Entropy created when trajectory is lost.

(If the trajectory carries information, that information is lost as well.)

A further complication may be required if we make the GHP distinction which implies reversible and irreversible sources for entropy.

This is a reduction of at least one and possibly two basic assumptions, depending on whether you count entropy arising out of trajectory as two or one. I would argue it's only one, accepting Boltzmann's explanation for entropy in kinetic systems.

A suggestion from Bennett[*] was to model a small chaotic system using a very few particles. The model here predicts that the information lost in trajectories would equal the entropy gained, except that one may have trouble describing the quantum homologue to entropy. Landauer asserted that energy expended in erasing information in a quantum computer increases the entropy of the system, but did not calculate it; apparently he thought it obvious.

A model is available[†] of an ideal gas of two particles among a few more other similar particles, tracking the entropy of the two particles and of the whole system as it is set in motion. This model predicts the result of an experiment in that system: start with the two particles in a high energy [momentum] state—the system entropy is relatively low at this point; set it in motion and allow the particles to interact briefly. After a few interactions, when entropy of the closed system has risen, reverse the trajectories. The two "interesting" particles will regain their initial higher energy [that is, the system will regain its original lower entropy]. Inserting standard quantum uncertainty into the position and momentum of the particles will render it impossible to regain the original state after a short time. In the first case, time's arrow will not be observable from within the system; in the second case obviously, it will. It would also be possible to discriminate effects due to computational truncation error [butterfly-wing instability] from those due to quantum uncertainty.

This all seems to reopen the question, frequently asked by the pioneers, is quantum mechanics a full description of reality? The answer seems to me to be no, which would be a great surprise. Hitherto, Quantum theory was thought to "coalesce" into classical models at large numbers, but that seems to break down at chaos, information and entropy. If information can be created in the macroscopic world that is lost in transition to the quantum

[*] Personal communication.

[†] M.M. Taylor, personal communication, has built an operating model of several billiard-ball universes.

world, even when it seems that it should be transmissible, it is a new twist on the double slit experiment where other information in the macro world (are there two slits open or one) cannot help but go into the quantum domain even when it seems that it shouldn't. It points up again, if it were needed, that relating quantum theory to chaos is critically important in developing theoretical physics. Without such unification, neither the quantum nor the classical model can be thought complete. There are efforts in that direction to be sure, but they have not yet gained acceptance‡.

We might decide to live with the conduction description above, as well as the turbulence description of convection, and to accept friction/conduction/turbulence-entropy-time as a working hypothesis. If we did, it would be well to develop experimental tests in order to confirm the information-theoretic implications. This game is worth the candle when you can reduce your assumption set by just one axiom, let alone by two.

The notion that time's irreversibility arises from the collapse of the possible quantum states is seductive. It begins to seem to me that such collapse is fundamental but no less mysterious than it ever was. Now, if a complete description of the collapse should involve the non-reversible weak interaction, that would tie it all up neatly. Of course, without real world experiments, studies, observation and confirmation--and often even with them--you never know if your elegant new idea is a Snark or a Boo...

ACKNOWLEDGMENT

Thanks for comments and suggestions to M.M. Taylor, DCIEM, and Giles Brassard, UQAM.

REFERENCES

1. Gleick J., 1987, *Chaos: The Making of a New Science*, Penguin Books, New York, NY.
2. Coveney P. and Highfield R., 1991, *The Arrow of Time: A Voyage through Science to Solve Time's Greatest Mystery*, Fawcett Columbine, NY.
3. Hawking S.W., 1998, *A Brief History of Time*, [Book. Originally pub 1988] Bantam Doubleday Dell, NY. [Film. 1992] Errol Morris, Director, VHS tape ASIN 630271SS05; DVD (in preparation) ASIN B00003CX99.
4. Zavartanik et alii, 1999, *This Proceedings*.

‡ While the idea of entropy is of great help in understanding classical mechanical systems, nobody has been able to find its analogue in quantum mechanics; therein lies the great unresolved mystery of quantum chaos (Gutzweiller, *op cit.*, 143.)

How was Time before time?

MARCO ALFANO
Centre for Communication Studies "Anghelos", Palermo, Italy

1. INTRODUCTION

Has the universe started to exist at some definite time? How did it happen? What existed "before" the universe?

Although difficult, an answer to these questions is important indeed not only for philosophers and religious but also for scientists, especially for those who envision a unity behind the huge variety of things (living and non) that make up the universe. It is important because it implies a different approach to the universe. There is a unity that "contains" everything and each thing is just a different aspect of this unity.

There are many scientists who look for the *theory of everything*. They try to "marry" the theory of relativity with the quantum theory in an attempt to provide single answers to questions coming from different disciplines such as the physics of particles and the astrophysics, i.e., from the infinitely small to the infinitely large.

Some of the most famous of these all-encompassing theories are *supersymmetry* (*susy*), *supergravity*, *superstrings* and *implicate order*[1]. In all these theories, as well as in their parent theories (relativity and quantum), time plays an important role even though it is mainly seen not as an entity of its own but rather as an aspect of the single field that permeates the whole universe. This field is often referred to as the *unified field* and has, among its aspects, the *space-time continuum* of the theory of relativity.

It is important to try to understand the properties and laws (assuming we can still talk of laws) which regulate this field if the general context in which time rises wants to be understood. Here an *holographic* view of the universe should be considered and *musical* laws should be applied.

Studies on the Structure of Time: From Physics to Psycho(path)ology
Edited by Buccheri *et al.*, Kluwer Academic / Plenum Publishers, New York, 2000.

This paper will first make some considerations on the unified field and the conditions that brought to the birth of the universe. It then will consider two theories of everything, i.e., *implicate order* and *superstrings* and how they deal with time. Finally, some personal considerations on time will be presented.

Note that in the context of this paper what is told about space will also hold true for time and vice-versa because they are considered just different aspects of the same space-time continuum.

Note, finally, that this is a speculative paper that tries to formulate some answers, but that mainly, as said above, wants to provide some suggestions to the reader for finding answers to the questions by himself or creating new and more precise questions in search of more satisfying answers.

2. ONCE UPON A TIME

Let us start with a story that tells us about the beginning of the universe.

Once upon a Time there was only pure Energy and there was nothing else. Yet this Energy did not exist because this Energy is all there was, and there was *nothing* else. And so, the Energy ... was not. For, in the absence of that *which is not*, that which is, is not.

This Energy knew of its existence conceptually but could not experience itself because there was no reference point outside itself.

This pure, unseen, unheard, unobserved and therefore unknown-by-anyone-else Energy chose to experience itself. Since there was no reference point outside itself, it realized it would have to use a reference point within. It reasoned that any portion of itself would necessarily have to be less than the whole and that if it thus simply divided Itself into portions, each portion, being less than the whole, could look back on the rest of itself and experience it.

And so this Energy divided itself becoming, in one "particular moment", that which is *this* and that which is *that*. For the first time, *this* and *that* existed, quite apart from each other. And still, both existed simultaneously. As did all that was *neither*. Thus three elements suddenly existed: that which is *here*, that which is *there* and that which is *neither here nor there* but which must exist for here and there to exist.

And thus, the physical universe was created because physicality was the only way for this Energy (and all its parts) to know experientially what it knew conceptually.

The created universe was a universe of *relativity*. This was a fundamental characteristic because it was impossible for anything to experience itself as it was until it had encountered what it was not. In fact, it is by which one is not that one is defined.

Thus, once upon innumerable times, the Energy knew Itself experientially...

We cannot, of course, "prove" if this story is "true" or not. Nevertheless, let us accept the story as it is for the moment. It is important because of the way it makes us to reason. What is important is the basic framework for our further reasoning.

It should be noted that Energy in this story has to be intended in a broader sense than we usually attribute to physical energy.

3. TIME AND TIME

When scientists began exploring the implications of Einstein's time for the universe as a whole, they made one of the most important discoveries in the history of human thought: that time, and hence all of the physical reality, must have had a definite origin in the past. If time is flexible and mutable, as Einstein demonstrated, then it is possible for time to come into existence and also to pass away again; there can be a beginning and an end of time. Today the origin of time is called the *big bang*. Religious people refer to it as the *creation*[2].

But what happened "before" the big bang? Has the universe being "always" existing or did it start in a specific Time?

Of course, literally speaking, it does not make sense to ask when the big bang happened, but rather how long ago the big bang happened. The singularity of the big bang (which is in any case a mathematical artifact) is defined to be a "boundary" to time not strictly part of time itself - not actually an event as such. The singularity bounds time in the past, implying that time has not endured forever. Thus, when encountering the word Time in the following text, it should not be intended as the time that existed before time but rather as a *state of existence* that transcends time and in parallel contains time.

As we will show in more details, we can consider Time as a *container* of times. This container is in reality a single field, the *space-time continuum*. What is important is the movement inside this field, (e.g., the spin for elementary particles). This movement causes space and time, as physical entities, to come to exist and change.

It is important to understand that the field is there, "always" and "anywhere", it is us who move from one part of the field to the other and, in doing so, we only focus on that specific part of the field, observing that specific time and that specific space. Thus, Time has not ended and never will have an end. It can be considered as *the eternal moment of now*.

Now, going back to the story of the previous section, we have that there was actually a Time before the beginning of time when there was no matter at all, just the purest form of energy. This purest form of energy forms a *unified field* that is the *potentiality* for all the energy and matter in the universe.

The unified field vibrates but it does not vibrate in space because it creates space. It does not vibrate in time either because it creates time. The unified field is the source of the space-time continuum. When the field vibrates, it creates waves, i.e., space-time events and becomes matter.

The whole universe now appears as a dynamic web of inseparable energy patterns. The universe is thus defined as a dynamic inseparable whole which always includes the observer in an essential way. If the universe is indeed composed of such a web, there is (logically) no such thing as a part. Thus all things (living and non) are no longer separated parts of a whole. They are a whole.

What are the characteristics of this unified field? Are there any works in physics that somehow refer to the existence of this unified field?

We believe there are and we now briefly present some interesting aspects of two important works in theoretical physics mainly aimed towards a theory of everything, namely the *implicate order* and *superstrings* theories. We describe how they assume the existence of a unified field and how they deal with space and time.

3.1 Implicate order - The holographic Time

While Einstein revolutionized the Newtonian concept of space and time, he nevertheless continued to assume that space (and hence time) is continuous. That is, the properties of space continue unchanged to smaller and smaller scales: Space can be divided and subdivided right down to the dimensionless point with all its properties changing smoothly from point to point[3].

Quantum theory shows that the infinite divisibility of space must be limited. The reason can be seen in Heisenberg's uncertainty principle which states that the energy confined within smaller and smaller regions of space-time becomes uncreasingly uncertain. In practice, Heisenberg's uncertainty principle tells us that the shorter a particular time interval, the more uncertain is the energy for the system during this interval. For very short intervals (equivalent to the collision times of high energy particles), this energy uncertainty may be very large, large enough to create a new particle. In practice, new particles are brought into existence for short time periods, only to die back and release their energy. It is as if quantum theory allows nature to borrow energy and then pay it back (fast enough) to create and then destroy particles provided that the sum of energy balances at the end of collision. The interactions that happen in this way are called *virtual interactions* and the particles created and destroyed are called *virtual particles*.

At the length scale of the elementary particles, these energy fluctuations do not pose a serious problem. But in much smaller regions, the borrowed energy can act to curve space-time (in Einstein's theory, both energy and matter distort the fabric of space-time). At around 10^{-33} cm, this borrowed energy is large enough to curve space-time right around itself. Thus, mini black holes can be created. Space (and hence time) loses its smooth continuous nature and breaks apart in a fluctuating foam[3].

David Bohm agreed that quantum theory is leading us toward a description of space that is topological rather than geometrical. Topology is concerned with boundaries, intersections and containment. Bohm felt that topological relationships rather than coordinate grids should form the basis of a new order that would be in harmony with the quantum theory.

Bohm pushed even further these ideas. He suggested that the explicit patterns and relationships we see around us may not be a manifestation of nature at its deepest level. Rather they are the explicate forms of an underlying *implicate* or *enfolded order*[4]. This implicate order exists in an unmanifested state (a state of potentiality) and is the foundation upon which all manifest reality rests. Bohm calls this manifest reality the *explicate* or *unfolded order*.

The implicate order has been compared to a holographic image in which information from many different parts of a scene is enfolded in a complex way onto the photographic plate (Fig 1). When the hologram is illuminated with the same kind of light used to create it (a laser beam), a whole holographical scene will appear and, depending on how we look at the hologram, we will see one part of the scene. An important property of a hologram is that every piece of the photographic plate is an exact representation of the whole scene and can be used to reconstruct the entire hologram.

The implicate order underlies space-time and matter so that objects which are normally thought of as being separate and distant become interlocking and mutually contained within each other. In practice, the totality of existence is enfolded within each region of space and time. Within the implicate order, for example, it is possible for distant events to be directly connected. Thus, one is led to a new notion of unbroken wholeness which denies the classical idea of analyzability of the world into separately and independently existing parts.

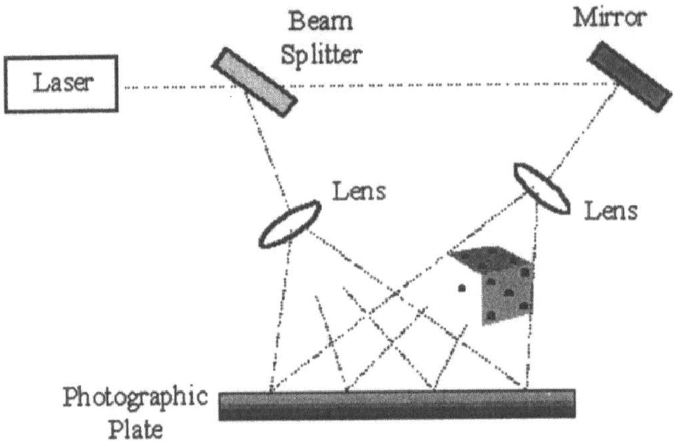

Figure 1. Creation of a hologram.

It should be added that what "carries" an implicate order (as the laser beam in the hologram that carries the information on the object to be reproduced) is the ***holomovement*** which is an unbroken and undivided totality. In certain cases, we can abstract particular aspects of the holomovement (e.g., light, electrons, etc.), but, more generally, all forms of the holomovement merge and are inseparable[4].

It should be noted that Karl Pribram, a brain researcher, has accumulated evidence that the brain's deep structure is essentially holographic[5]. He states that research from many laboratories by sophisticated analysis of temporal and/or spatial frequencies demonstrates that the brain structures sight, hearing, taste, smell and touch holographically. The information is distributed throughout the system so that each fragment can produce the information of the whole. Pribram uses the hologram model to describe not only the brain but the universe as well. He states that the brain employs a holographic process to connect to a holographic domain that transcends time and space.

3.2 Superstrings - The music of Time

The theory of *superstrings*[3] proposes that the elementary particles cannot be reduced to featureless points but are essentially one-dimensional objects, *strings* that rotate and vibrate. Not only matter finds its origins in these one-dimensional objects of incredibly minute size (10^{-33} cm), but strings themselves are irreducibly linked to an underlying space. Indeed space-time itself must at some level be created out of superstrings.

But how can it be that the ultimate entities of our universe are strings?

In order to answer this question we have to go back of two thousand years to the time of Pythagoras. He and his followers believed in the importance of numbers and in the deep link between numbers and music. Simple numerical ratios, as the ones applied to a string of a musical instrument (e.g., 1:2, 2:3, 3:4) to generate the most consonant intervals (e.g., octave, fifth, fourth), would produce harmony. This numerical harmony not only would be present in music but would also extend, by *resonance*, throughout the universe. The music produced in the universe (by the motion of stars and planets) was known as *the music of the spheres*.

In the superstrings theory, what appear to be elementary particles are in fact the notes or the localized vibrations (fundamental plus harmonics) of quantum mechanical strings. In essence, the superstrings theory proposes that the origin of all particles lies in tiny quantum strings that in their vibrations and rotations (quantum strings differently from normal strings can also rotate) produce a pattern of *quantum notes*. Although a string does not have a mass itself, its quantum notes do because they are localized energy levels. Since we know that energy has an associated mass, given by $E=mc^2$, so each excitation or note of a string has an associated mass.

The same should also apply to space and time. A string has not a space and a time associated to it (although it has been postulated that a string has a dimension of 10^{-33} cm.) whereas its quantum notes do.

The vibrational quanta are widely spaced like the rungs on a ladder of energy. Corresponding to each of these rugs, there is a smaller ladder whose closely spaced rungs correspond to the string's rotations (Fig 2). The result is like having a series of musical scales that represent the vibrations and rotations of the strings.

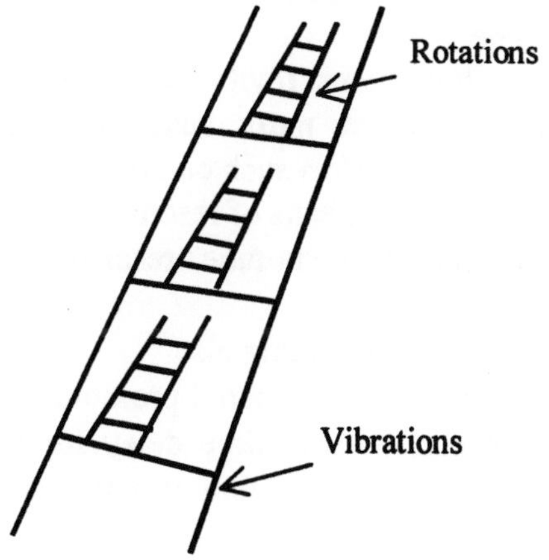

Figure 2. Energy levels corresponding to vibrations and rotations of strings.

It should be noted that in recent studies[6] astronomers have been using techniques drawn from the analysis of music in order to study the harmonic components of energy distributions at very large scales and derive how galaxies form into progressively larger groupings. In these studies, once again the term *music of the spheres* has been used.

4. THE FIELD

In the previous section, we have briefly presented some particular aspects of two theories that assume the existence of a unified field underlying the physical reality. The holographic model assumes a single unbroken holomovement which "carries" the implicate order and is the foundation upon which all manifest reality rests. The quantum strings can be seen as the components of a single field that generates (through vibrations and rotations of quantum strings) all elementary particles and hence all physical reality. But, are these two theories dealing with two different fields?

If carefully examined, we see that, although focusing on different aspects, the two theories present many common points. Going even further, we can say that these theories are just focusing on different aspects of the same *Field*. What are then the main characteristics of this Field?

By summing up and, at the same time, abstracting the features of the two theories presented above, we have that the Field presents the following characteristics:

1. It is made of vibrations at all frequencies

Vibrations are not only physical and frequencies should not be considered as the inverse of (physical) time because, at this level, it does not make sense to talk of time.

2. It vibrates and creates holograms. It vibrates and creates music

The Field vibrates and creates mass, space, time and all the physical reality. It is the movement that makes such creation possible.

3. Vibrations are transmitted by means of resonance

Things (living and non) that are "in tune" resonate exchanging vibrations with no time delay.

4. Everything is connected to everything else

Connections are not dependent on spatial proximity or on time. An event in one location immediately, without time delay for communication (i.e., faster than light and beyond Einstein's relativity) affects everything else.

5. Each piece contains the whole

Each part of us contains the whole pattern of us. This can be expressed in the physical world in our genes. In them, each cell contains our whole genetic makeup.

In order to see the hologram, we must illuminate it with a laser beam of the same kind of the one that was used to create the hologram. The more *synchronous* the used laser beam is to the original one, the more of the whole scene we will be able to generate.

6. Each aspect of the universe is individual and not identical to any other aspect

The interactions that make an aspect of the universe to emerge are always different, so an aspect will always be different from other aspects. As an example, let us consider light. Light, in each individual observation, will present itself either as a particle or as wave but never as a particle and a wave at the same time.

7. The whole is greater than the sum of the parts

The Field is *synergistic*, i.e., the simultaneous actions of separate elements have a greater total effect than the sum of the individual effects. As a consequence, each system of elements exists within a system greater than itself, which also exists within a system greater than itself, and so on.

Each of the above statements would require a deep discussion by its own on the consequences it implies about the way of approaching and understanding our universe, but this is outside the scope of this work. Nevertheless, an interesting question that spontaneously arises is: Can we somehow "observe" the characteristics of the Field discussed above?

Perhaps the reader could try to give an answer on his/her own based on his/her subjective experience or reformulate the statements in a way that makes more sense to him/her, thus providing more satisfying answers.

Another important property of the Field is the following:

8. Time is also holographic

Each aspect exists everywhere, all the times, and always (i.e., both at all times and in all times). Each moment is whole and complete and coexists in a knowledgeable relationship with all other moments.

Thus the Field exists in Time, the so called *eternal moment of now* and contains all times in an interconnected whole. It is us who, interacting with the Field, experience a specific time. Notice, however, that events that at first sight are thought to be independent often show, by means of *synchronicity*, a deeper and more subtle link.

Let us conclude this work with an analogy. Let us think of a music score where different notes appear separate and distinct, yet each existing "simultaneously with the other" (Fig 3).

Figure 3. Music score.

If it is true that usually notes are played left to right (and top to down) along the staff, nevertheless we can always play notes picking them randomly from the score. Moreover, if we eliminate the staff between notes we somehow also eliminate the temporal order of note playing (Fig 4).

Figure 4. Notes without a reference staff.

If we now consider notes as times, we see times all existing "simultaneously" with the others, yet separated. It is just us who look at one specific time (at the time).

We can look at the different time moments from left to right, but we can also look at them, for example, right to left (backward time) or even jumping between different time moments (time travel).

The score is in front of us, the times are there. It is up to us choosing in which way we want to move between different times.

The score is the Time. It is the potentiality for all notes (times).

5. CONCLUSIONS

This paper has discussed time in the general framework of Time. The characteristics of the Field underlying all the physical reality have been presented and the holographical and musical nature of Time has been discussed.

The next questions that naturally come from the above discussion are: Can we perceive Time? How do we perceive it?

It is difficult to answer these questions. If Time has a holographical nature, the more we are able to "illuminate" an event in time with a "coherent" light (as a laser beam does with a hologram), the more we will be able to experience Time. In practical terms, this could entail, for example, to "see" the synchronicity among seemingly unrelated events.

On the other hand, when we interact with a single event such as, for example, in a physical observation, we will observe a specific time that is not absolute but relative (as Einstein taught us) to that specific interaction (for example, how the observer and the observed move with respect to each other).

A final consideration (or rather provocation) is that while physical time is considered to be objective and psychological time is considered to be subjective, from the above discussion we have that psychological time looks at Time and physical time is relative and then mutable. Thus, which one is "really" objective and which one is subjective?

REFERENCES

1. Lederman L., 1996, *The God Particle*.
2. Davies P., 1995, *About Time*, Penguin Books.
3. Peat F.D., 1988, *Superstrings and the Search for the Theory of Everything*, Contemporary Books.
4. Bohm D., 1980, *Wholeness and the Implicate Order*, Routledge.
5. Wilber K. (ed.), 1982, *The Holographic Paradigm and othet Paradoxes*, Shambala.
6. Landy S.D., 1999, Mapping the Universe, *Scientific American*, June.

Three philosophical points about time

GIANNI RIGAMONTI
Dipartimento di Matematica, Via Archirafi 34, 90123 Palermo, Italy

1. INTRODUCTION

There are three points about time and the way it usually is discussed today that have, in my opinion, an outstanding philosophical significance. The first is whether time has to be thought of as something uniform and isotropic, or there can be an essential difference between past and future; the second is whether it can be seen as a set of instants and, more generally, whether continua can be seen as point sets (where "point" has a very general meaning I will explain presently); the third is whether all we need to describe temporal events are instantaneous snapshots, or something else, not reducible to such snapshots, is also indispensable. Each section of the present work is devoted to a discussion of one of these topics.

1. In modern physics time is conceived not as an arrow, but as a straight line. This is especially obvious in relativity theory, where standard reference frames are four-dimensional and one of their dimensions is time, whose mathematical treatment is not significantly different from those of length, width, and depth. Now, a straight line, whether Euclidean or non-Euclidean, is a *homogeneous* entity whose points and portions share a common, unvarying nature. It is true that, when we have a metric for it, some of its points will have a positive abscissa and some a negative one: but this is not an essential difference, for it depends on our choice of a zero-point, and such a choice is entirely arbitrary. So if time really is just one more dimension and really has the same structure as a straight line, there is no essential difference between future and past: times to come and times gone by are the same sort of thing, just like points right and left of zero on a straight line. And that is, indeed, the perspective

prevailing in modern physics: the same ontological status is ascribed both to past (or present) and future.

Now, this way of thinking involves forgetting about (or not being familiar with) an issue which cannot be reasonably neglected, namely that the ontological similarity of past and present can be (and has often been) challenged with serious arguments. In fact, according to a long-standing tradition the present and the past are real, for present things are happening now, past ones have happened, and neither can be made, as it were, to un-happen. If it is now the case that P, we cannot claim it is not now the case that P; and if it has been the case that P, we cannot claim it has not been the case that P. If we did, we would incur into a contradiction. But no such contradiction exists, in general, when we say *either* that it will be, *or* that it will not be the case that P. Indeed, *for some sorts of* P'*s* (to be defined presently) there is nothing impossible *neither* in claiming that it *will* be the case that P, *nor* in claiming that it will *not* be the case. For concerning the present and the past, we are constrained: either P or not-P is *given*, and that is why asserting, respectively, that not-P or that P involves a contradiction. But how can "It will be the case that P", or "It will not be the case that P", be given, if neither P, nor not-P is occurring, or has occurred? There is nothing for our statements to contradict here.

The answer to this argument is well known: in a sense the future *is* given, since, for all P's, either P or not-P is bound to happen. Being bound to happen is something as unmodifiable as having happened or happening now; it is true it is inaccessible to us in a way past and present occurrences are not, but this is a difference just in our vision, not in things as they really are. *Per se*, the future is just as determined as the present and the past.

A traditional way of stating this contrast involves an old-fashioned term, that of *future contingent.* Future contingents are non-predetermined future events, whose taking place or not taking place is not decided in advance. Determinism denies future contingents, non-determinism admits them. More exactly, non-determinism (in its most reasonable and authoritative variants) admits that *some* future events (such as, for example, the occurrence of an eclipse at a certain date) will necessarily take place, but also maintains that certain other ones (typically, our choices) may take place *and* may not take place at some future time. For determinists, this twofold possibility is illusory and only reflects our ignorance. I don't know whether I will buy a new pair of shoes within this month (and nobody else does), but my not *knowing* it does not mean that my *doing* (or not doing) so is not predetermined; in fact, despite my present ignorance, it is as given and unchangeable as a past or present event. For non-determinists, on the opposite, it really is a future contingent, i. e. something whose occurrence and non-occurrence are both possible today.

This is a philosophical problem and has to be decided, or declared unsolvable, on philosophical grounds. Four positions are possible (and all four have been defended in the history of thought): (I) everything is determined, (II) nothing is determined, (III) some future events are determined and some are not, (IV) we don't know. Personally, I strongly sympathize with position (III); but, of course, not everybody shares my sympathies. There is, however, a methodological point everybody agrees with, formally at least: if discussions are to be serious, problems cannot be swept under the rug. Now, when time is described taking for granted (as it is often done in physics) that all of its parts have the same sort of being and, more specifically, that the future is as real as the past (the so-called "realism about the future"), this *is* sweeping a big philosophical issue under the rug. I would not call that serious.

2. There is another controversial point about today's prevailing notion of time, namely the well-known discrepancy between physical and bio-psychological time. But the difference is not simply that the latter has an arrow the former lacks;[1] there also is another and perhaps deeper point, i. e., whether time can be seen as a set of instants. This problem, however, is just a species of a wider genus, namely whether continua at large can be seen as "point" sets, where by "point" in a *k*-dimensional continuum (1_*k*<_) I mean an ordered *k*-tuple of real numbers. A physical interpretation can be superposed upon this numerical framework, and different continua correspond to different interpretations. To illustrate: if *k*=1, in the standard view the elements of a one-dimensional continuum, i. e. the *real numbers* it consists of, can be interpreted as distances from a given zero-point (in which case we have length), or as instants (and we have time), or as temperatures, weights, and so on. Time is, then, a set of instants, space a set of points in the ordinary, geometrical sense, etc. This is standard in modern physics *and mathematics*.

   But, received as it is, this view can be countered with a classical and very robust argument. Many continua - such as length, surface, (ordinary) space, time, weight, or mass - are additive. This means

   (a) that every object in such a continuum has a measure according to a parameter associated with the continuum itself (every object in space has a volume, every event in time has a duration, and so on);

[1] It is true something is beginning to change in physics (see Zavrtanik's contribution); but, in my opinion, just locally. The global perspective of physicists upon time stille sees irreversibility as a mystery.

(b) that when such an object is divided into parts, whether finitely or infinitely many, each of these parts has in turn a measure according to the same parameter, *and the sum of the measures of the parts is equal to the measure of the whole* (additivity).

More exactly, additivity presupposes an *exhaustive* and *pairwise disjoint* division of a continuum, i. e., a division into parts such that every point of the continuum belongs to some part and none belongs to two distinct parts. But if parts are not pointlike, i. e., have in turn parts, a relation of *contiguity* can exist between each of them and some other ones; where if two parts A and B are contiguous no distance exists between them (more precisely: for every real number α there exist $x \in A$, $y \in B$ such that the distance between x and y is smaller than α) but they don't overlap. Just imagine, to visualize, a line AC and a midpoint B in it: there is no distance between AB and BC, but they don't overlap (of course, the midpoint can belong either to AB, or to BC. It doesn't matter which we choose).

This "no distance, no overlap" relation can be realized *physically* in all *physical* additive continua; and when it is, a whole is produced whose measure is the sum of the measures of the parts. Take for example length: if we have two sticks AB and CD whose lengths are respectively l_1 and l_2, we can join them end to end, letting B (or A) touch C (or D), and get a new stick whose length is $l_1 + l_2$. Or take two bodies A and B, weighing respectively w_1 and w_2: we pile them upon one another and get a new object weighing $w_1 + w_2$.

Now, this is impossible with points, and not just because they all have zero length (and breadth, and depth) and zeros, when added together, always yield a zero sum - which is the standard argument - but also for another, deeper reason: there cannot be a no-distance, no-overlap relation between points. If there is no distance between two of them, then in fact they are one; for they don't simply "touch", and don't simply overlap either, but merge and are indistinguishable; and if there is no overlap they cannot be in touch, but a distance must exist between them. With points, only two relation are possible: separation, or complete superposition. There can be neither simple touch, nor partial overlap (let alone inclusion, with point A being a proper part of point B). For touch and overlap presuppose parts: "A touches B" means that no distance exists between a certain *part* of A and a certain *part* of B, not between *all* of A and *all* of B (that would be superposition, not touch). And "A partially overlaps with B" means that a part of A (not all of it) coincides with a part of B (not with all of it). But points have no parts: so if two of them touch, all of one touches all of the other, and they coincide. And the same goes for overlap, of course. There is no difference between touch, overlap, and sameness. Moreover, if you let objects having a non-zero measure touch, you make a bigger object of the same sort thereby (a wider surface with surfaces, a longer duration with durations, and so on), but if you let any number of points touch, you always get just *one* point.

This is a time-honored, Aristotelian argument; but neither has it ever been refuted, nor do I see how it could be. So it only remains to decide whether, whenever a continuum is additive, additivity is an essential property for it; since if it is, *that* continuum cannot be a point set, such sets being obviously non-additive. I believe this decision is up to philosophy and philosophical preferences; thus, seeing continua at large as point sets implies an usually unconscious, but no less real for that, philosophical preference. And that preference is not the only possible one.

3. Time is additive. So there is a difficulty in conceiving time, too, as a point set - or, more specifically, as a set of instants. But here this general difficulty is compounded with a specific one: for if we take time to be a set of instants, then any temporal event is a set of snapshots, one for each instant of its duration, and a description of a temporal event can only be an (obviously finite) selection from among such snapshots. This is, for example, exactly what modern physics does: it is only interested in states of the world, or of some smaller system, at times t, t', t'' or whatever, where all these "times" are instants. And outside physics, though certainly influenced by it, there are philosophers, such as Quine, who believe that every empirical statement comes with an implicit or explicit time qualification and clearly show, through all the examples they give, to have only instantaneous qualifications in mind. In this sense, today's prevailing ontology of time is monistic: time is a set of instants and there only are instants within time. No other sort of objects exists there - not, for example, intervals, understood as something primitive, unreducible to instants: for in the standard view intervals *are* reducible, being nothing but sets of instants themselves.

   This standard treatment of time begins to look ungrounded once we challenge the idea that continua are point sets, and time, accordingly, a set of instants - or rather *the* set of *all* instants. But we also have a more specific reason to conclude that a dualistic ontology of time, admitting instants *and* durations not reducible to instants, is to be preferred: there are things that cannot be well explained in terms of temporal cross-sections or snapshots, while in terms of a synthetic unity presupposing an unreducible *flow* of time they can. Among such things, the most important is by far speech: every time we utter a sentence, such as "Please let us sit down and have lunch", the uttering would be incomprehensible without this synthetic unity. For a sentence (a) is a succession: having constituent parts, and a linear ordering of these parts, is an essential aspect of its nature. But (b) it is not just a succession: for if a goose honks, and one second later a crow croaks, and two seconds later a cat mews, and three seconds later a dog barks, and four seconds later a horse neighs, and five seconds later a pig grunts, and six seconds later a little bird chirps, this also is a succession, but it has not the synthetic unity a sentence has. A sentence-uttering is something we do not simply *hear* across time: we *understand* it across time, i. e. as an event (i) having a duration, (ii) divided into certain well-defined temporal subsections, (iii) bound to last, or having lasted, or both, beyond any of these subsections. But this means that a sentence is a temporal manifold being understood as *one* thing; and it also means something stronger. For the unity of this manifold is such, that its constituents do not make any sense if taken individually, insulated from each other; this is true of letters, of syllables, and also of words, if we take Frege's principle that a word only makes sense inside a sentence seriously.

A sentence, then, cannot be understood as a succession of temporal cross-sections or snapshots. In fact, it cannot even be understood as a succession of sounds, each with its own duration (since there can be, for example, long and short syllables). We can understand the honk-croak-mew-bark-neigh-grunt-chirp like that: not a sentence.

Music functions in much the same way. We perceive tunes, not notes; this is so true, that almost everybody can recognize a tune he or she has already heard, while very few people can recognize single notes. To tell whether a single note is an A, a B or whatever, you need a so-called absolute ear, and absolute ear is most uncommon, even among professional musicians; while, supposing you have previously heard tune so-and-so, you need no rare quality to be able to say "This is tune so-and-so" when you hear it again. You simply need normality. So music is typically perceived as a synthetic unity of notes across time; for that is a tune. It is like a sentence, not like a honk-croak-mew-bark-neigh-grunt-chirp.

There are, then, some very important things we can only make sense of if we take the flow of time as something essential, not identifiable with a (possibly continuous) set of instants. From this viewpoint a dualistic ontology of time, describing durations and instants as unreducibly different things, might have some merit.

REFERENCES

1. Aristotle, 1998, *Physics*, Ross (ed.), Oxford University Press.

2. Aristotle, 1988, *Prior and Posterior Analytics*, Ross (ed.), Oxford University Press.
3. Cantor G., 1932, *Gesammelte Werke mathematischen und philosophischen Inhalts*, Zermelo (ed.), Olms Verlag, Hildesheim.
4. Dedekind R., 1926, "Continuità e numeri irrazionali", *Essenza e significato dei numeri*, Oscar Zariski Stock, Roma.
5. Einstein A., 1968, *Relatività*, Esposizione divulgativa Boringhieri, Torino.
6. Frege G., 1967, *Logica e aritmetica*, Corrado Mangione Boringhieri, Torino.
7. Newton I., 1964, *Philosophiae naturalis principia mathematica* (Italian translation), Pala (ed.), UTET, Torino.

Author index

Subject index

Appendix to the chapter "Algebraic Geometry: A Tool for Resolving the Enigma of Time?" by Metod Saniga.

English translation of the excerpts

Sect. 4.1

Excerpt in Italian: p.149

"*Time is standing still for me*, I believe. It is perhaps only a few moments that I have been so bad. I look at a clock and I have the impression, if I look at it again, that an *enormous period of time* has passed, as if hours would have passed instead only a few minutes. It seems to me that the duration of time is enormous. *Time does not pass any longer, I look at the clock but its hands are always at the same position, they no longer move, they no longer go on; then I check if the clock came to a halt, I see that it works, but the hands are standing still.* I do not think about my past, I remember it but I do not think about it too much. When I am so bad, *I never think about my past.* Nothing enters my mind, nothing... I did not manage to think about anything. I did *not* manage to see *anything in my future. The present does not exist for me* when I am so bad... *the past does not exist, the future does not exist.*"

1st excerpt in German (starting "...Das seltsame war..."): p.149

"...The strangest thing was that every once in a while my normal time-awareness, as far as these figures were concerned, got totally lost; *time was no longer a stream, which flew away and whose flux could have been measured, but it was rather similar to a sea, which as a whole stood still and which was in itself only a chaotic and utter jumble. I was no longer able to understand the continuous becoming of the figures as a sequence in a certain time direction, but sometimes the colours and forms flew into an indescribable jumble, as if the previously alternating figures were now experienced all simultaneously.* Had I previously seen these figures in a constant motion, so now it was only a colorful and inexpressible manifold there in which I was not able to perceive any motion anymore. When I totally sank into the show of the figures, it happened every now and then that I also sank *into this time-still-standing, where the succession was transformed into a still standing present.* Not only am I now not able to formulate these interruptions of the normal time experience, I am also almost unable to imagine my experience of them any more. When I tore myself away from these figures and violently turned myself to the outer world, this anomalous time experience was no longer here, but this disturbance of the sense of time found its expression in a form of illusion that an *immense long* time must have passed since my last waking-up."

2nd excerpt in German (starting "Ich merke wohl..."): p.149

"I sure do notice the passing of time but can't experience it. I know that tomorrow will be another day again but don't feel it approaching. I can estimate the past in terms of years but I don't have any connection to it anymore. *The time standstill is infinite, I live in a constant eternity. I see the clocks turn but for me time does not flow...*

Everything lies in one line, there are no differences of depth anymore... Everything is like a firm plane."

Sect. 4.2

Excerpt in German: p.150

"I woke up in a whole different world in which the puzzle of the world was solved extremely easily in the form of a different space. I was amazed at the wonder of this different space and this amazement concealed my judgement, this space is totally distinct from the one we all know. It had different dimensions, everything contained everything else. I was this space and this space was me. The outer space was a part of this space, I was in the outer space and the outer space was in me...

Anyway, I didn't experience *time, time of the outer space and eons* until the second phase of this dream. In the cosmic flow of time you saw worlds coming to existence, blooming like flowers, actually existing and then disappearing. It was an endless game. *If you looked back into the past, you saw eons, if you looked forward into the future there were eons stretching into the eternity and this eternity was contained in the point of the present.* One was situated in a state of being in which the "will-be" and the "vanishing" were already included, and this "being" was my consciousness. It contained it all..."

Sect. 4.3

1st excerpt in German (starting "Das Denken ist..."): p.152

"Thinking is different, it has no style any more. *What is with the future? One cannot get hold of it. One can speak about the present and past, yet one can no more imagine them.* One cannot depict them, one must go down. Everything is like a question mark...

It pulls me back, well, where to? To where it comes from, there, where it was before. *It enters the past. It is that kind of a feeling as if you had to fall back. This is the disappearing, the vanishing of things. The time slips into the past*, the walls are fallen apart. Everything was so solid before. It is as if it were so close as to be grabbed, as if you had to pull it back again: *Is that the time? Shifted way back!*

In the morning when I wake up, well how I can say, the vanishing is again here: it affects me deeply! Do I know where I am? Oh, yes. *But the vanishing, that there is no time here, and where one is seized by time and what was yesterday! There it goes on inside of me, always farther backward, – but where? Time is in collapse.*"

2nd excerpt in German (starting "Ich bleibe stehen..."): p. 152

"I stop still, *I am being thrown back into the past* by words that are being said in the hall. But this all is self-evident, it must be that way! *There is no present anymore, there is only this stated being related to the past, which is more than a feeling, it goes through and through...*

Is there any future at all? Before, the future existed for me but now it is shrinking more and more. The past is so very obtrusive, it throws itself over me; it pulls me back...

I am like a burning arrow that you hurl before you; then it stops, falls back, and is finally extinguished as if in a space empty of air. By this I want to say *that there is no future and I am thrown back... Strange thoughts enter my mind and drive me off into the past.* It is terrible. It goes through and through!..."

3rd excerpt in German (starting "Ich kann nichts mehr..."): p.153

"I cannot anticipate anything at all, *as if there were no future anymore. I used to think about what I would do next week or what I would buy myself in the summer, I cannot do that now. I always believe, it all stops now and there is altogether no tomorrow anymore.* I feel like a doll, which is without life, to which a new life first has to be given. *I cannot think of the future – I cannot anticipate*." (And what about *the past*?) "*Actually, I think about it very intensively*. It is very painful for me to think that those wonderful moments are gone." (And *the present*?) "*I don't think all too much about the present. I just take it as it comes, the day goes by*. I always used to study where this all comes from." (What is happening with your thinking?) "Now, I can already think, only forward, it is not simple, it is, in fact, as if it were cut off."

Sect. 4.4

1st excerpt in German (starting "Gestern am Mittag..."): p.154

"Yesterday at noon, when the meal was being served, I looked at the clock: why did no one else? But there was something strange about it. For the clock did not help me any more and did not have anything to say to me any more. How was I going to relate to the clock? *I felt as if I had been put back, as if something of the past returned, so to speak, toward me, as if I were going on a journey. It was as if at 11:30 a.m. it was 11:00 a.m. again, but not only time repeated itself again, but all that had happened for me during that time as well*. In fact, all of this is much too profound for me to express. In the middle of all this something happened which did not seem to belong here. *Suddenly, it was not only 11:00 a.m. again, but a time which passed a long time before was there and there inside* – have I already told you about a nut in a great, hard shell? It was like that again: *in the middle of time I was coming from the past towards myself*. It was dreadful. I told myself that perhaps the clock had been set back, the orderlies wanted to play a stupid trick with the clock. I tried to envisage time as usual, but I could not do it; and then came a feeling of horrible expectation that *I could be sucked up into the past, or that the past would overcome me and flow over me*. It was disquieting that someone could play with time like that, somewhat daemonic..."

Excerpt in Italian: p. 154

"*...I got the impression that time was flowing backward; I felt that time proceeded in the opposite direction*, I had just this extraordinary sensation, indeed... the most important sensation at that moment was, *time in the opposite direction...* The perception was so real that I looked at a clock and, I do not know how, I had the impression that the clock confirmed this feeling, although I was not able to discern the motion of its hands..."

2nd excerpt in German (starting "Bei meinem..."): p. 154

"As I suddenly broke down I had this feeling inside me that time had completely flown away. After those three weeks in a sick-camp, I had this feeling that the clock hands run idle, that they do not have any hold. This was my sudden feeling. I did not find, so to speak, any hold of a clock and of life anymore, I experienced a dreadful psychological breakdown. I do not know the reason why I especially became conscious of the clock. *At the same time, I had this feeling that the clock hands run backward...*There is only one piece left, so to speak, and that stands still. I could not believe that time really did advance, and that is why I thought that the clock hands did not have any hold and ran idle...*As I worked and worked again, and worried and did not manage anything, I simply had this feeling that everything around us (including us) goes back... In my sickness I simply did not come along and then I had this delusion inside me that time runs backward...*I did not know what was what anymore, and I always thought that I was losing my mind. *I always thought that the clock hands run the wrong way round, that they are without any meaning. I just stood-up in the sick-camp and looked at the clock – and it came to me then at once: well, what is this, time runs the wrong way round?!...I saw, of course, that the hands moved forward, but, as I could not believe it, I kept thinking that in reality the clock runs backward...*"

Sect. 5

1st excerpt in German (starting "...Im Treppensteigen..."): p. 158

"...While walking upstairs, a sudden and as if nailed-down picture of this moment, the momentary view of Dr. M., Dr. St. and myself in space, attracted my attention. This repeated itself on different stairs. At the top of the stairway there seemed to be *no continuity of time at all, the whole course of events was only a mess of separate situations without any connection. And these situations, in case of active work, could later have been connected in the same way in which one can observe a celluloid film. Yet at the same time these situations – in both experiencing and a direct reproduction of the happening afterwards – carried the character of the independent and disconnected.* A strange next-to-each-other-ness, not a one-after-the-other-ness; *they have no position in time*, time has no sense here..."

2nd excerpt in German (starting "...Die Unterscheidung von Gegenwart..."): p. 158

"...The distinction of the present and the future is not cancelled out as the patient still speaks about both dimensions, yet *the line between the actual present and the only maybe-possible and unreal future becomes swaying and possible to cross. Both dimensions encapsulate and overlap each other without a steady transition.* The future fuses with the present and vice versa and experiencing acquires a flickering twilight character *which is radically distinguished from how a healthy person anticipates the future in day-dreams and the like... The edge between the present and the past is swaying as well.* At the same time and in a totally different way, the past is included in and fuses with the events of the present as well as usually the present is a part of the past. *There is a kind of condensation of time; the present is not distinguished amidst the continuous-steady flow of the past any more*, but at the same time the present is not filled with something past as it usually is with normal people; in this case it overlaps.

In these hallucinations that were totally aimed at the future it was, from our point of view, the emphasis now clearly shifted on that which was to happen in the future. And which sensually and as if in person broke into the present, *in the course of which both these levels of time experience got quicker and encapsulated without any continuous transition*. Through the phenomenon of "déja-vu," which became part of the real situation, the borders to the past also turned out to be swaying: *one dimension of the past, which was not integrated with the rest of the past, pushed its way into the present in a discontinuous manner...*"

ACKNOWLEDGEMENTS

I would like to express my cordial thanks to Miss Daniela Veverkova and Mr. Peter Hahman for translating into English all the excerpts quoted from papers written in German: the only exception is the first account in Sect. 4.1, whose translation was taken from an English edition of Ref. 46, pp. 284-5. My warm thanks are due also to my friend Dr. Rosolino Buccheri for the corresponding translation of a couple of excerpts in Italian.

MIX
Papier aus verantwortungsvollen Quellen
Paper from responsible sources
FSC® C105338

If you have any concerns about our products,
you can contact us on
ProductSafety@springernature.com

In case Publisher is established outside the EU,
the EU authorized representative is:
Springer Nature Customer Service Center GmbH
Europaplatz 3, 69115 Heidelberg, Germany

Printed by Libri Plureos GmbH
in Hamburg, Germany